Studies in Computational Intelligence 453

Editor-in-Chief

Prof. Janusz Kacprzyk
Systems Research Institute
Polish Academy of Sciences
ul. Newelska 6
01-447 Warsaw
Poland
E-mail: kacprzyk@ibspan.waw.pl

T0122616

For further volumes:
http://www.springer.com/series/7092

Xia Wang and Wolfgang A. Halang

Discovery and Selection of Semantic Web Services

 Springer

Authors
Dr. Xia Wang
GFaI e.V.
Berlin

Prof. Dr. Dr. Wolfgang A. Halang
Lehrstuhl für Informationstechnik
Fernuniversität
Hagen

ISSN 1860-949X
ISBN 978-3-642-42760-2
DOI 10.1007/978-3-642-33938-7
Springer Heidelberg New York Dordrecht London

e-ISSN 1860-9503
ISBN 978-3-642-33938-7 (eBook)

Printed on acid-free paper

Springer is part of Springer Science+Business Media (www.springer.com)

To Xia's parents for their love, understanding and support

Abstract

The next generation of web search engines not only promises to be able to search for semantically related information dispersed over different web pages, but also for semantic services which are functional entities performing certain tasks. This functionality depends on the maturity of the technologies for the Semantic Web (SW) and Semantic Web Service (SWS). With the development of the semantic web and semantic web services, the discovery of semantic services becomes the key problem. To address it, this book presents a systematic method for web service discovery providing optimal search results under specific user service requirements.

Semantic web service discovery is, after all, a matchmaking process, which semantically measures the similarity of a web service's description with user requirements. So far, service discovery has not been addressed clearly and comprehensively. The quality and results provided by current approaches are still incomplete and often erroneous, because either keyword searching or simple functional matchmaking is used, only. Service discovery is a difficult problem not only due to its complexity, but also due to the limitations of pertaining research results available. Approaches borrowed from information retrieval, knowledge representation and discovery, data mining, artificial intelligence, ontology engineering and computing logics are considered to be adapted in order to solve this problem.

After an extensive and detailed investigation of related work, this book identifies four problems not addressed by current solutions: (1) A clear web service model suitable for selection is lacking, because there is too much excrescent information interfering with the selection process and impairing its operational efficiency, if service selection is directly based on service descriptions. (2) No common domain ontologies exist for certain kinds of web services and, thus, service providers are inclined to use their own service ontologies in describing their services. The resulting diversity of ontologies presently prevailing renders discovery, selection and interoperation of services very difficult, if not impossible. (3) Without a uniform domain ontology serving as knowledge base, the current ontology-based service discovery approach is very weak. It lacks rich semantic information to support matchmaking of services. (4) Service selection based on Quality-of-Service (QoS) criteria is currently not well developed and efficient. Syntactic measurement and evaluation of all quality metrics of services are missing, rendering the discovery of services impossible.

Addressing these four problems, in this book the following contributions are presented: (1) A novel service model is proposed, which is independent of semantic service description models, and clearly defines all elements necessary for service discovery and selection. Such a service model takes service selection as its gist and improves its efficiency. Corresponding selection algorithms and their implementation as components of the eXtended Semantically Enabled Service-oriented Architecture (xSESA) in the Web Service Modeling Environment (WSMX) are presented in detail. (2) A general approach for building application ontologies is presented, of which many applications of semantic web services, e.g. discovery, composition and mediation, can benefit. With application ontologies thus built, service discovery can be carried out in the same way as with single domain ontologies, and the mediation problem between service ontologies turns out to be solved. Then, an ontology-based approach to improve service discovery is proposed and thoroughly validated. (3) Using the service model presented, a service selection approach oriented at QoS-criteria is proposed. It normalises diverse qualities of a service in their respective metrics and employs a service selection algorithm based on soundness.

A scalable experimental platform, named Web Services Application Ontology (wsao) builder, has been developed to evaluate the work above. It systematically demonstrates the feasibility of the architecture proposed to discover semantic services. Experiments reveal that precision and recall of service selection based on application ontologies are significantly improved by about 15.1% and 14.7%, respectively.

Acknowledgements. Dipl.-Ing. Jutta Düring's contribution to this book in form of drawing most of the figures contained is highly appreciated.

Contents

Acronyms

AI	Artificial Intelligence
AO	Application Ontology
BPEL4WS	Business Process Execution Language for Web Services
DAML	DARPA Agent Mark-up Language
DARPA	Defence Advanced Research Projects Agency
DC	Discovery Component
DERI	Digital Enterprise Research Institute
DL	Description Logic
FCA	Formal Concept Analysis
HTTP	HyperText Transfer Protocol
IMsg	Input Message
IOPE	Inputs, Outputs, Pre-conditions and Effects
IRS	Internet Reasoning Service
OIL	Ontology Interchange Language
OMsg	Output Message
OWL	Web Ontology Language
OWL-S	Web Ontology Language for Service
QoS	Quality of Service
RDF	Resource Description Framework
RuleML	Rule Mark-up Language
SAWSDL	Semantic Annotations for WSDL
SC	Selection Component
SMTP	Simple Mail Transfer Protocol
SO	Service Ontology
SOAP	Simple Object Access Protocol
SP	Service Provider
SR	Service Requester
SW	Semantic Web
SWRL	Semantic Web Rule Language
SWS	Semantic Web Service

tModels	Technical Models
UDDI	Universal Description Discovery and Integration
UML	Unified Modeling Language
WS	Web Service
WSAO	Web Service Application Ontology
WSCI	Web Service Choreography Interface
WSD	Word Sense Disambiguation
WSDL	Web Service Description Language
WSMF	Web Service Modeling Framework
WSML	Web Service Modeling Language
WSMO	Web Service Modeling Ontology
WSMX	Web Service Modeling eXecution environment
WSPDS	Web Services Peer-to-peer Discovery Service
WWW	World Wide Web
XML	eXtensible Mark-up Language
xSESA	eXtended Semantically Enabled Service-oriented Architecture

Chapter 1
Introduction

Today we are in an era of information explosion and overload. The amount of information available in the World Wide Web is exploding, and the web's infrastructure acts as a huge storage of electronically linked web documents. Although the web made it possible to access and share information and services anywhere in the Internet, human users are increasingly dissatisfied with being able to browse static information (e.g. news, weather, hotel information), only, and having to manually filter desired information out of thousands of documents returned by the current search engines, e.g. *Google* or *Yahoo!*. Instead, they desire the next generation of the web to be intelligent and to provide electronic services, but not static data, and to support them in automatically performing both simple and complex tasks.

In other words, intelligent and interactive web-based services are required. A service is actually an abstract resource that represents the capability of tasks. Automatically searching for a service on the Internet is, therefore, far more difficult than just mining for a piece of text information as provided by the contemporary web search engines, and it is like looking for a needle in a haystack. Achieving this goal becomes, however, possible by means of semantic web service technologies. Hence, let us begin a brief history of these technologies' development.

Supported by the set of Web Service technologies [83], there were gradually thousands upon thousands of web services published, which are accessible functional entities to perform certain tasks. The capabilities of web services are initially and syntactically described in their respective service descriptions using the *Web Service Description Language* (WSDL) [33]. The idea is that users can find web services by matching published web service descriptions with their requirements. The discovery of web services is, however, not an easy task — actually, it is a great challenge.

To enhance the popularity of the semantic web technology [14] and to facilitate intelligent service applications, it was suggested to integrate web services into the semantic web in the form of *Semantic Web Services (SWS)* [107] by endowing web services with semantics. So far, the semantics of web services have not been defined clearly. To the best of our knowledge, a service's semantics means its knowledge base, which is normally represented by an ontology and logic, and which is

X. Wang and W.A. Halang: Discovery and Selection of Semantic Web Services, SCI 453, pp. 1–8.
springerlink.com

described in a machine-readable form to allow for agreement between itself and other service entities in order to achieve interoperation. Moreover, semantic information enriches web services with expressibility of real behaviour and reasoning capabilities enabling their understanding and handling by machines.

The perspective of automatically and intelligently discovering and invoking services via the web to achieve any tasks will promote people's life greatly and have important significance. For instance, to arrange one's holidays, it is not necessary to personally visit travel agencies, but one can enter one's holiday requirements into a web-based search engine, which then looks for a set of — in a certain sense — best services including booking flights, reserving accommodation and planning entertainment. A more complex example is that of an international freight agent responsible for accepting orders and arranging shipping all over the world, while the actual shipment will be delegated to other delivery companies. By using web service technologies, the agent's tasks can distributed to a service engine to automatically discover and compose heterogeneous services.

Although semantic web services are such a promising technology, it still requires a deal of effort to make them suitable for daily routine use. There are many open challenges in the field of semantic web services, such as web service description, discovery, selection, composition, mediation (at data, process and service levels) or invocation (especially in dynamic environments). Basing on recent contributions found in the literature, especially with respect to describing semantic services, this book focuses on how to discover apt semantic web services. This is certainly a key issue, and its solution will contribute to the settlement of the other open problems in service mediation as well as composition.

In the remainder of this chapter, we first state the problems encountered and, then, clearly specify the research questions to be addressed by this book. Then, we indicate the scientific contributions which will be made as body of the book and, finally, an overview of the content of this work will be given.

1.1 Problem Statement

Initially, web service descriptions were uniformly defined in WSDL. This is an abstract representation language, which syntactically structures the information of a web service, consisting of the service's name, its operations, parameters, messages types and optional binding protocols, aiming to advertise the service and to let users know how to invoke it. Obviously, WSDL was developed from the viewpoint of service invocation, i.e. when a machine or a system decides to invoke a web service, then by its WSDL description one can clearly know by which protocol or with what kind of messages to invoke it. But how to decide which web service is the one needed is an issue of web service discovery.

The discovery of web services focuses on measuring the similarity between the descriptions of published web services and given service requirements. If the similarity of two service descriptions is simply and literally measured by matching their *.wsdl* documents (just structurally matching service names, operation names, inputs and outputs), the result of web service discovery is likely to be imprecise or even erroneous due to lacking semantic support [176, 124]. If a user is searching for a flight-booking service, for instance, without the prior knowledge that "flight" is used synonymously with "airline" (although these two concepts are not identical lexically), services searching for airlines will not be found as candidate services.

There is a similar problem with *Universal Description Discovery and Integration (UDDI)* [165, 168], as service discovery through UDDI ultimately returns a service location, a key defined in its technical model (*tModel*[1]), which is a service interface and defines the technical specification for a service in UDDI). Due to lacking semantic description, UDDI basically uses keywords to match services. Hence, the results are inclined to be erroneous or just wrong.

Therefore, first of all a specific representation language to express the semantics of web services is necessary, which should have the capabilities to describe services with semantic information and to explicitly reason about description similarity. Secondly, corresponding and more powerful matching algorithms need be defined. In the course of this book, our discussions of service discovery will always emanate from these two aspects.

On the one hand, there are currently two mainstream web ontology languages employed to describe services: the *Web Ontology Language for Service (OWL-S)* [104] supported by the *World Wide Web Consortium* (W3C[2]) and the *Defense Advanced Research Projects Agency* (DARPA[3]), and the *Web Services Modeling Ontology* (WSMO) [24] developed by the *Digital Enterprise Research Institute* (DERI[4]). From the perspective of service discovery and selection, we briefly summarise these two languages as follows.

- OWL-S is, per se, a semantic mark-up language for web services basing on the Web Ontology Language (OWL) [38]. It has three main parts: *service profile*, *process model* and *grounding*. Only the service profile aiming to advertise and discover services presents the capabilities of services, including non-functional and functional properties as well as a few Quality-of-Service (QoS) attributes. The semantic reasoning capabilities are, however, somehow weak for service matching purposes.
- WSMO is an ontology and conceptual framework to completely describe web services and their most related aspects in terms of the Web Service Modeling Framework (WSMF) [51]. Although WSMO describes the capabilities of web services, it covers too wide a range of service scenarios to improve service interoperability, and it mainly focuses on solving the integration problem. Also, the implementation mechanism for choreography and orchestration is still under development.
- OWL-S and WSMO adopt similar views on the service ontology but rely on different logics, viz. OWL/SWRL (Semantic Web Rule Language) and

---

[1] http://www.uddi.org/pubs/uddi_v3.htm

[2] http://www.w3c.org

[3] http://www.darpa.mil

[4] http://www.deri.organdhttp://www.wsmx.org

WSML-Core/WSML-DL/WSML-Flight/WSML-Rule/WSML-Full (Web Service Modeling Language[5]), respectively.

Conclusion. From our point of view, WSMO is the better choice when it comes to representation and discovery of services. Therefore, all work presented in this book is to be considered in the WSMO framework.

On the other hand, the current matching algorithms can roughly be divided into four categories on the basis of their methods' focus (although their content might have intersections), namely *capability-based* and *QoS-based* (at object level), *description-logic-based* and *ontology-based* (at meta-level) ones. They are elaborated in Chapter 3. To summarise, the following problems are commonly encountered in course of applying them.

- The algorithms used by [69, 31, 176, 74] to match service descriptions are either based on keyword searching or matching entire service descriptions as in documents. Since the description of a semantic service is a well structured mark-up document and not just a simple bag of words, the current algorithms are insufficient to address the problem of service discovery.
- The algorithms used by [162, 124] are based on the hypothesis that a single domain ontology is shared and used in the descriptions of semantic web services and service requirements. Current applications use their own ontologies, however, and do not consider mapping and similarity of heterogeneous ontologies.
- The algorithms used by [162, 124, 89, 29] yield approximate results when expressing the similarity of two services, only. As in [124], for instance, four matching degrees are defined, viz. *exact*, *plug-in* (in case $outA$ subsumes $outR$)[6], *subsume* (in case $outR$ subsumes $outA$) and *fail*. As this approach fails to provide quantitative service measures, it is hard to know which web service is the one matching best.
- Reasoning capabilities based on service representations [59, 124, 65] are still weak. Supplementarily, several rule languages, such as the *Semantic Web Rule Language* (SWRL) by *daml.org* or the *Rule Markup Language* (RuleML) by *ruleml.org*, are under development.

Summary. The analysis above looked from two sides, web service representation and matching algorithms, and revealed the current problems of semantic web service discovery. Thus, considering the facts above, the approach taken in this work is (1) to select WSMO and WSDL as languages for web service description, and (2) to propose an effective matching algorithm to discover services by measuring service similarity based on both capabilities and quality properties of services.

1.2 Research Questions

For reasons of clarity, here we specify the research questions to be addressed in this book. All in all, the central question investigated is:

---

[5] WSML, http://www.wsmo.org/TR/d16/d16.1/v0.21/

[6] *outA* is the output of advertised web services and *outR* is the user's requirement.

How to select a semantic web service that matches given service requirements best, by evaluating the similarity of service requirements against web service descriptions?

This theme gives rise to the following concrete research questions.

1. *What is an appropriate service model for SWS discovery?*
   Aiming to improve the state of the art in discovering semantic web services, the first prerequisite is to define a service model supporting service discovery and selection. Such a service model is needed to solve the essential problem of discovery. In this service model, only the necessary service features are formalised and defined to ensure the validity of the results of service selection.
2. *How to build an application domain ontology for a kind of SWS?*
   The second prerequisite is to address how to build an application domain ontology for a kind of semantic web services, in order to resolve the mismatch between heterogeneous service ontologies and to improve service discovery by much richer semantics. Using this single application ontology during the discovery process of semantic services, service similarity can be measured exactly. Moreover, such an application ontology is also a good knowledge base for any mediation at the data level, process level and service level.
3. *How to measure concept similarity in an application ontology built?*
   Web services are using different ontologies to describe the services they provide. Even the same or similar kinds of web services can be represented by heterogeneous service ontologies. Therefore, the similarity of ontological concepts among web services is the primary issue to service discovery. By resolving the mismatch of service ontologies in the application ontology built, service matching can really be improved.
4. *How to measure service similarity by evaluating service qualities?*
   To discover a web service matching some user requirements best, a search engine does not only match service functions, but also, and more importantly, it should consider the qualities of web services. Service qualities are, however, diverse, complex and complicated to be defined and measured. A proper approach to a combined evaluation of the service qualities is still lacking, because many service qualities have different scales, metrics or features. Although it is difficult to select services by their qualities, it is important to employ qualities in service discovery.

1.3 Contributions

The work reported in this book addresses the questions indicated in the preceding section. Our contributions focus on the development of methods to discover and select semantic web services. Specifically, these contributions fall into four major categories:

1. *Identifying a service model for service discovery and selection*
   Despite the large body of work available in the area of service definition and on the frameworks for semantic web services, little work was directed towards

identifying a service model appropriate for service discovery and selection purposes. In order to compensate for this deficiency, we define a semantic service model, which formalises all of the features necessary for service selection. Under this service model, selection algorithms are described in detail. The service model has been applied in real web service scenarios to illustrate its performance of service discovery and selection. A concrete implementation of the model is investigated in Section 4.3. It is developed as a service component of the extended semantically enabled service-oriented architecture (xSESA) (cp. Section 4.3.1) in WSMX.

2. *Method to build an application ontology for semantic web services*
   While ontologies play a very important rôle in any application of semantic web services, little attention has been devoted so far to the ways of automating their construction. This is not only because building an ontology is a time-consuming and boring task, but also because there are no guidelines and no tools supporting ontology building in the context of semantic web services. In this book, we present a basic approach to build application ontologies by merging existing service ontologies, which gives a general idea on how to build ontologies for more universal applications.

   In order to build an application ontology in SWS (cp. Chapter 5), first, we formally define it in the WSML language, and represent it as a semantic net with multiple concept relations. Secondly, we take WordNet as standard knowledge base to perform word sense disambiguation. Then, a WordNet concept extraction algorithm is employed to formalise service ontologies. Finally, a WordNet-based ontology merging algorithm is applied to generate the application ontology by merging service ontologies. During the rule-based building process of the application ontology, conflicts between ontologies are considered. Moreover, several experiments have been made to evaluate their effectiveness (cp. Chapter 8).

3. *Web services matching based on service ontology similarity*
   The most important goal of our work is to improve the mechanisms of service discovery. Just from the ontology point of view, little attention has been given to mediate heterogeneous service ontologies in eliminating their mismatches in order to discover services matching best. This book extends the ontology definition of WSMO and represents it as a semantic net with weighted multiple concept relations. In such a semantic net, ontological concept similarity is calculated (cp. Chapter 6). Finally, a corresponding service similarity algorithm is proposed by considering ontological concept similarity. This part of the work has been evaluated and compared with similar approaches, yielding very encouraging results (cp. Chapter 8).

4. *Service selection by synthetically evaluating service qualities*
   Although the QoS-based approach of service discovery has received broad attention in the literature, there is still a lack of work on measuring, in an integrated way, all different qualities in a discovery case. Therefore, after formally defining a quality ontology in WSMO, we present a QoS-based selection model for web services. A novel QoS-based selection algorithm is then proposed, which uses a

normalisation algorithm oriented at optimal value ranges and, finally, leads to a fair quantitative assessment (cp. Chapter 7).

1.4 Structure of This Book

This book is structured into four parts with nine chapters covering the four research questions posted above.

1. *Introduction and problem statement*
   After a brief introduction of the background and significance of research on semantic web services, in the present first chapter we stated the problems of selecting semantic web services, surveyed the related work very briefly and listed the four research questions to be dealt with in this book. Moreover, the contributions were then summarised and the book structure is presented here.

2. *Research context and related work*
   Chapter 2 focuses on the introduction of the research context and the preliminary knowledge about the semantic web and semantic web services. Using real scenarios, step by step we introduce web services and their related technologies (WSDL and UDDI), then the semantic web and semantically enabled web service. The chapter ends with a detailed analysis of two kinds of ontologies and their relations with semantic web services.

   A detailed overview of related work on the technology of semantic service discovery is provided in Chapter 3. According to the order of technical development, we discuss the strengths and weaknesses of the different approaches to discover web services and semantic ones. Moreover, ontology building and merging in the context of semantic web services is investigated with putting forward our points.

3. *Semantic web service discovery and selection*
   This part is the proper body of this book and consists of four chapters corresponding to our four contributions. Aiming to address the research Question 1, in Chapter 4 we abstractly identify a service model supporting semantic service discovery and selection, which can be implemented in any of the current SWS frameworks. We then propose discovery and selection algorithms appropriate for this service model. For validation purposes, this service model has been implemented in form of two service components of WSMX and plugged into the *xSESA* architecture.

   Chapters 5 and 6 present, respectively, the solutions to research Question 2 and 3 stated in Section 1.2, which are how to build an application ontology for semantic services, and how to use this application ontology to calculate concept similarity to improve service selection. Detailed algorithms are finally evaluated on our *WSAO* studio experimental platform and prove the soundness of our ideas.

   Addressing research Question 4, in Chapter 7 the proposed QoS-based approach to service selection is presented, which solves the problem of how to evaluate service similarity by measuring all different features of service quality.

Chapter 8 evaluates our work on plenty of collected real data step by step. We compare our results with the ones of related work and prove the effectiveness of our methods.

4. *Conclusion and future directions*

Finally, in Chapter 9 we state our conclusions by summarising our contributions, and point out three potential directions for future research, which can benefit from our achievements.

Chapter 2
Semantic Web Services and Ontologies

This chapter summarises the scientific context and the state of research in the area addressed in this book. We present a brief overview of web services, semantic web services and ontology-related technologies. To foster easy understanding, we use real examples from some concrete scenarios to illustrate how these technologies work, and which limitations and problems remain.

2.1 Web Service Technologies

What is a web service? In 2002 the W3C Web Services Architecture Working Group defined a web service as a *"software system designed to support interoperable machine-to-machine interaction over a network. It has an interface described in a machine-processable format (specifically WSDL). Other systems interact with the web service in a manner prescribed by its description using SOAP messages, typically conveyed using HTTP with an XML serialisation in conjunction with other web-related standards"*.

To put it simple, a web service is a software component which can be accessed based on its machine-readable service description via a web interface. Web service technologies actually provide standard means of interoperability between different software applications, running on a variety of platforms and frameworks. Thanks to the use of a set of XML-based technologies, a web service itself is indeed a self-describing, self-contained, modular application that can automatically be published, located and invoked by machines.

Fig. 2.1 gives an overview of the web service standards, all based on XML. By using Unicode[1] and Namespace[2], the Extensible Mark-up Language (XML) [23] provides the foundation of data structures and document formats, and facilitates information systems to share their data via the Internet. Thus, any kind of data can

---

[1] http://www.w3.org/TR/unicode-xml/
[2] http://www.w3.org/TR/REC-xml-names/

X. Wang and W.A. Halang: Discovery and Selection of Semantic Web Services, SCI 453, pp. 9–26.
springerlink.com   © Springer-Verlag Berlin Heidelberg 2013

Fig. 2.1 Web service standards by [1]

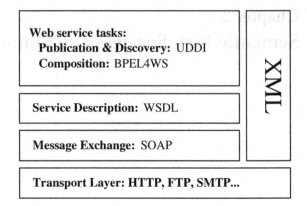

be exchanged between web services (e.g. textual, semi-structured or structured) as long as it is embedded in an XML format.

In the transport layer, several underlying protocols, such as HTTP, FTP and SMTP, are choices to be bound by messages according to the Simple Object Access Protocol (SOAP) [22] during communication. For instance, the Hypertext Transfer Protocol (HTTP) [119] is applied to transfer interlinked text documents (hypertext, identified on the web using Uniform Resource Identifiers (URIs)), and HTTP's GET and POST are the two methods usually employed to request or respond to SOAP messages via the Internet.

The interface of a web service is described using the Web Service Description Language (WSDL), and its message exchange is encoded in the SOAP and transported by Internet protocols. Tasks like publication, discovery or composition can be performed with web services.

Example scenario. We refer to the W3C travel reservation use-case to define our scenario[3]. A travel agent offers to clients the ability to book a complete travel package consisting of airplane/train/bus tickets, hotels, car rental, excursions etc. Its service providers, i.e. airlines, bus companies or hotels, are providing web services allowing to query their offerings and perform reservations, and credit card companies provide services to guarantee payments made by clients. For instance, an airline normally has the following functions provided by its web service.

```
service(
  PortType:TicketAgentSoap(
     op:listFlights(
         IMsg(listFlightsRequest),OMsg(listFlightsResponse))
     op:reserveFlights(
         IMsg(reserveFlightRequest),OMsg(reserveFlightResponse))
     op:cancleFlights(
         IMsg(referenceCode),OMsg(cancelResponse))
  )
)
```

---

[3] http://www.w3.org/2002/04/17-ws-usecase

Service description. The example above shows a schematic representation of the WSDL file associated to a ticket service[4]. Industry considerably supports WSDL, also increasingly by tools such as WSDL generators and editors. As an XML-based language, it is a machine-processable, structured and standardised way to describe web interfaces of services. In WSDL, a service is seen as a collection of network endpoints which operate on messages. The example service provides one **port**:*TicketAgentSoap*. This port groups together three **operations** that deal with ticket searching (by *listFlights*), reservation (by *reserveFlights*) and cancelation (by *cancleFlights*). Each operation has an input (IMsg) and an output (OMsg) message.

A **message** has a name and a set of **parts** of certain types. Parts represent input/ouput parameters depending on whether they are declared in the input or the output message, e.g. the following representation of *message:listFlightsRequest*. The **type** of a part can be any XMLSchema data type or a previously defined complex type (for brevity, an example of type is omitted here.)

```
message:listFlightsRequest(
        part: name="depart" type="xsd:dateTime"
        part: name="origin" type="xsd:string"
        part: name="destination" type="xsd:string"
    )
```

In a nutshell, a WSDL document has two major parts. First, the *abstract interface* of the service specifies the data types, messages and portTypes with the corresponding operations (which refer to previously defined messages). Second, an *implementation part* binds the abstract interface to concrete network protocols and message formats (e.g. SOAP, HTTP).

Web services architecture. A typical web service architecture is illustrated in Fig. 2.2. A *service provider* defines and publishes the WSDL description of its service to a *service registry* (i.e. a UDDI), a place maintaining the description information of all registered web services. Subsequently, *service requesters* can inspect the UDDI and locate/discover web services that are of interest to them. Using the information provided by the WSDL descriptions, they can directly invoke the corresponding web services.

The centralised approach of the above web service model is not the only way to publish web services. They can also be published by a distributed environment, such as a peer-to-peer or a grid network.

The main purpose of describing and publishing a web service is that it is enabled to be found and invoked to carry out some tasks. In the following section we briefly focus on two tasks of web services.

Service discovery. Broadly speaking, this is the act of locating a machine-processable description of a web service which meets certain requirements (defined in a *query*). It is based on matching the requirements against the published service advertisements, and finding one or sometimes a set of web services. Although there

---

[4] http://schemas.airlines.org/TicketAgent.wsdl

Fig. 2.2 Web service model by [83]

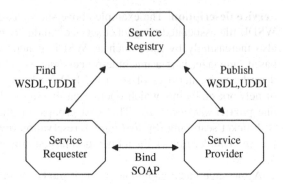

are many different interpretations of web service discovery, agreeing with [19] we deem that it consists of a set of subactivities, and that one should consider to differentiate between service discovery at service design time and at run time (cp. Chapter 4). As a discovery example, in our scenario a query for a flight ticket could be specified as:

```
OneWayTicket(
        r: departing= "Dublin"
        r: arriving= "Munich"
        r: departureDate= "2008-12-14"
        r: departureTime= "late afternoon"
    ) // r means "require"
```

By matching this query against web service descriptions, a search engine will discover any candidate airline. Assuming twenty airline services are found having flights from Dublin to Munich on 14 December, with only five flights in the afternoon, then this search engine will return the five services matched in accordance with the price in descending order.

Service composition. In fact, several web services can be composed to achieve more complex functionalities. Web service composition is the way to automatically compose a set of atomic web services with a control and data flow to fulfill a business goal. For instance, if a client is offered a special discount in a five star hotel on certain days, then he/she has to base on these dates and the hotel's location the entire travel arrangement, including flight bookings and car hire. A possible composition policy is illustrated in Fig. 2.3.

As WSDL specifies the static and syntactical descriptions of single services, only, a language for the specification of web service flows is needed in order to enable service composition. The Business Process Execution Language for Web Services (BPEL4WS[5]) or the Web Service Choreography Interface (WSCI[6]) are such specification languages. The workflow technology and planning methods of artificial intelligence are the means currently used for web service composition, as indicated in the detailed surveys [142] and [17].

---

[5] http://download.boulder.ibm.com/ibmdl/pub/software/dw/specs/ws-bpel/ws-bpel.pdf

[6] http://www.w3.org/TR/wsci/

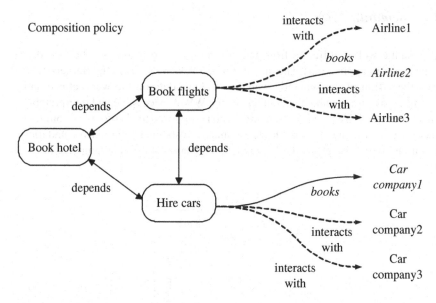

Fig. 2.3 Composing travel-related web services

Limitations of web service technologies. Although web services can hide any implementation details from the end users to achieve the goal of cross-language and cross-platform operation by relying on the XML-based standards WSDL, SOAP and UDDI, there are still many limitations preventing fully automated interoperability. First of all, human labour is still required in defining or interpreting WSDL files [126]. Indeed, WSDL specifies the functionality of web service only at the syntactic level. Although such descriptions can automatically be parsed and invoked by machines, the interpretation of their meaning is left to human programmers.

To support dynamic service discovery or composition, both WSDL and UDDI are not sufficient. The capabilities of describing web services are currently limited to their syntax and taxonomy, i.e. still far away from being machine-understandable. They lack expressiveness for logics, not even to think about pre-conditions for and effects of service operations [129]. To summarise, real computer-processable semantics are needed, which include the semantics of service properties, capabilities, interfaces and effects [1].

2.2 Semantic Web Services

At the same time, the development of the Semantic Web (SW) appears encouraging for web services. As [107] said, the semantic web community addresses the limitations of the current web service technology by augmenting service descriptions with a semantic layer in order to achieve their automatic discovery, composition, monitoring and execution.

2.2.1 Semantic Web

The semantic web is different from the current web and to be used beyond display purposes in applications. The inventor of the WWW, Sir Timothy Berners-Lee, coined the vision of the semantic web as, "to have data on the web defined and linked in a way and to make the contents of the WWW accessible and interpretable by machines" [14]. By [158], the semantic web is supposed to (1) provide a common syntax for machine-understandable statements, (2) establish common vocabularies as in an ontology, (3) agree on a logical language and (4) use that language to exchange proofs[7].

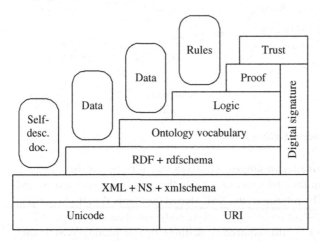

Fig. 2.4 The language layers of the semantic web

Fig. 2.4 illustrates the language layers of the semantic web suggested by Berner-Lee[8] and discussed in detail by, e.g., [127]. As already mentioned in Section 2.1, the URIs provide a standard way to refer to entities, and Unicode is a standard to exchange symbols. XML/XML Schema with namespaces provides a simple, very flexible text format to enable the exchange of a wide variety of data on the web [23]. These two layers are widely accepted and used on the web.

RDF is seen as the first layer related to the semantic web. It is a general-purpose language to represent information on the web by using three types of entities (resources, properties and statements). In RDF a statement is defines as an *object-attribute-value* triple to express a simple knowledge item. RDF Schema (e.g. Dublin Core metadata[9]) is a specification to describe how to use RDF to describe RDF vocabularies [37].

The next layer is the ontology vocabulary, which represents a shared understanding of some domain of interest by formally defined domain concepts and concept

---

[7] A proof is a justification statement accompanied with the results provided by the semantic web to convince a user to accept and trust the results [131].

[8] http://www.w3c.org/DesignIssues/Semantic.html

[9] The Dublin Core Metadata Initiative, http://dublincore.org/

relations. The ontology vocabulary can facilitate exchange and understanding of knowledge data. While, as representation of a knowledge base, only with *logic* an ontology can let the vision of the semantic web become reality. That is, with certain knowledge rules or axioms expressed in a logical programming language, a machine may reason/infer new knowledge from the stated and formalised knowledge. The remaining layers of *proof* and *trust* are to be provided to check the validity of statements in the semantic web by their creators.

Ontology Web Language. Obviously, the implementation of the semantic web depends on the web ontology language used and its expressiveness for logic. After content mark-up languages inspired by artificial intelligence such as OIL[10] and DAML+OIL[11], the Web Ontology Language (OWL), a successor of DAML+OIL, exploits the results of some two decades of research into Description Logic (DL) [117]. Its specification was accepted as standard by the W3C Web Ontology Language Group in 2004.

Basing on Description Logic and frame-based logic, OWL has rich classes, properties and axioms to explicitly represent concepts and concept interrelationships. Moreover, OWL has three increasingly expressive sublanguages: OWL-Lite, OWL-DL and OWL-Full. These languages have well defined semantics and enable the mark-up and manipulation of complex taxonomic and logical relations between entities on the web. Supported by a set of tools[12], OWL can automatically be processed by applications. For instance, "luxuryHotel" is a subclass of "Hotel", and if restricted to be a three star hotel, it can be expressed by OWL-Lite as follows[13].

```
<owl:Class rdf:ID="LuxuryHotel">
    <rdfs:subClassOf>
        <owl:Class rdf:about="#Hotel"/>
    </rdfs:subClassOf>
    <rdfs:subClassOf>
        <owl:Restriction>
            <owl:hasValue rdf:resource="#ThreeStarRating"/>
            <owl:onProperty>
                <owl:ObjectProperty rdf:about="#hasRating"/>
            </owl:onProperty>
        </owl:Restriction>
    </rdfs:subClassOf>
</owl:Class>
```

2.2.2 Search Engines for the Semantic Web

With more expressive web description languages, the semantic web is to support more efficient information or knowledge discovery, automation and integration.

---

[10] http://www.ontoknowledge.org/oil/TR/primitives.html

[11] http://www.daml.org/2000/10/daml-oil

[12] http://www.w3.org/2001/sw/WebOnt/impls

[13] Extracted from http://protege.cim3.net/file/pub/ontologies/travel/travel.owl

Although search engines for the semantic web are presently still in the research and development stages, the providers of search engines such as Google, Yahoo or Microsoft are trying to launch such products, e.g. Yahoo! SearchMonkey[14], Google's new "Knowledge Graph"[15] and Hakia[16] [166].

Based on keyword searching and state-of-the-art algorithms, contemporary search engines can efficiently answer topical queries [95]. They fall short, however, in answering intelligent user queries due to their results' dependence on information available in web pages. With this approach, they either show inaccurate results or they show accurate, but possibly unreliable ones. Owing to many limitations, there still remains a long way to go for really semantic search engines [85], particularly for those searching for non-textual web services.

2.2.3 SPARQL as Query Language for the Semantic Web

As one of the key technologies of the semantic web, the recursively named *SPARQL Protocol And RDF Query Language* [133] is a query language to retrieve and manipulate data stored in RDF format. Since most existing technologies of the semantic web service are based on RDF, in a May 2006 interview[17] Sir Timothy Berners-Lee expressed his expectation that "SPARQL will make a huge difference".

Information in the web is represented by RDF using triples containing subjects, predicates and objects. Correspondingly, most forms of SPARQL queries contain sets of triple patterns, which differ from RDF triples only by subject, predicate and object possibly being variables. As SPARQL allows queries not only to consist of triple patterns, but also of conjunctions, disjunctions and optional patterns, one is able to formulate very complex queries in it. A query "to find all landlocked countries with a population greater than 15 million people, with the country of highest population listed first", for instance, can be written as follows.

> **PREFIX** type: <http://dbpedia.org/class/yago/>
> **PREFIX** prop: <http://dbpedia.org/property/>
> **SELECT** ?country_name ?population **WHERE** {
>   ?country a type: LandlockedCountries;
>   rdfs:label ?country_name;
>   prop:populationEstimate ?population.
>   **FILTER** (?population > 15000000) **& &**
>   langMatches (lang(?country_name),"EN").
>   } **ORDER BY DESC**(?population)

---

[14] http://developer.yahoo.com/searchmonkey/siteowner.html

[15] Which evolved in part out of Googles acquisition of Metaweb in 2010, now claiming to understand 500 million entities and 3.5 billion attributes and connections; more at http://allthingsd.com/20120516/google-gets-semantic-launches-knowledge-graph-in-english-starting-today/

[16] A general-purpose semantic search engine that searches structured text like Wikipedia.

[17] http://en.mercopress.com/2006/05/24/web-inventor-internet-is-ready-for-big-leap

Such a query can be distributed to multiple SPARQL endpoints, i.e. services accepting SPARQL queries and returning results. These results are then gathered and combined, a procedure known as federated query. Employing SPARQL makes it possible to retrieve and manipulate any data stored in RDF-based datasets. Since existing web ontology languages, such as OWL and WSML, are RDF-based, SPARQL queries can thus be extended to provide answers involving ontologies, e.g. OWL-S ontologies [154].

2.2.4 Semantically Enabled Web Services

Bringing semantics to web services enables Semantic Web Services (SWS) [107]. This technology can be deemed as bond between semantic web and web services technologies and, thus, SWS has the potential to set the trend from syntax and statics to semantics and dynamics (see Fig. 2.5). The vision of semantic web services by [108] is to describe web service capabilities and content in an unambiguous computer-interpretable language, and to improve the quality and robustness of operations such as service discovery, invocation, execution and composition.

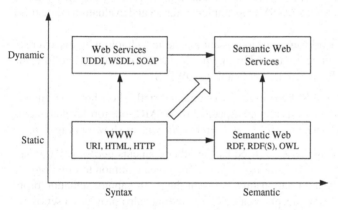

Fig. 2.5 Semantic-web-enabled web services

As described by [19], **"service semantics** *represent the machine-readable meaning of the task performed by a service. It is characterised by the pre-conditions and effects of the service execution"*. Then, the semantics of a service can simply be regarded as the behaviour expected when interacting with the service. To make service behaviour machine-understandable and use-apparent, a specific ontology language for creating semantic mark-up of web services is an urgent need. After its predecessor DAML-S[18]), OWL-S[19] was quickly accepted as specification of a semantic service mark-up language by W3C in 2004. Later on, several other semantic web service ontologies were launched from different views of SWSs, such as WSMO[20],

---

[18] http://www.daml.org/services/daml-s/0.7/daml-s.html
[19] http://www.w3.org/Submission/OWL-S/
[20] Web Service Modeling Ontology, http://www.wsmo.org/

SWSO[21] and WSDL-S[22] (for overviews and comparisons cp. [86] and [27]). More recently, WSMO bases on the principle of the Web Service Modeling Framework (WSMF) [51] to define various aspects related to semantic web services, including *ontologies*, *goals*, *web services* and *mediators*. The goal and mediation aspects of web services are rarely taken into account by the other web service ontologies. Several important semantic web service frameworks such as OWL-S, WSMO, IRS [114] or METEOR-S [128] consider service discovery, interoperation, composition and invocation.

Our research context. The work presented in this book was performed during the time when OWL-S was developed and WSMO was in the starting phase. Although having some overlaps with OWL-S, WSMO brings several important additions to OWL-S (for a detailed analysis and comparison of them cp. [86]). Moreover, the gradually completed specifications (including WSML[23] and WSMX[24]) and plenty of implementation tools strongly support WSMO, aiming to make semantic web services a reality. Currently, these tools consist of an implementation API (*wsmo4j-0.6.1*[25]), an editor (*wsmo studio-0.7.3*[26]), an execution environment (*wsmx-0.5*[27]), an administration tool (*wsmt-1.4.1*[28]) and different reasoners (such as *wsml2reasoner-0.6.2*[29], *DL reasoner v.0.1*[30] and *Rule reasoner v.0.1*[31]). By all of them it was suggested to take WSMO as implementation and evaluation context for most of this research.

Nevertheless, our work supposes to present a generic technology to discover semantic services without being limited by any particular specification. Therefore, the results of our research are equally applicable to OWL-S.

WSMO. In the sequel, WSMO is used as example to briefly show how to bring semantics to web services and its related aspects. For WSMO four top-level elements are defined, *ontologies*, *web services*, *goals* and *mediators*, as shown in Fig. 2.6.

Ontologies as knowledge base are used by the other elements *goals*, *web services* and *mediators*. In WSMO, *ontologies* define an agreed common terminology by providing concepts and relationships between them. To capture semantic properties of relations and concepts, generally an ontology also provides a set of axioms, which are expressions in some logical language. The usage of ontologies by a web service will be illustrated by examples in Section 2.4.2.

---

[21] Semantic Web Services Ontology, http://www.daml.org/services/swsf/1.0/swso/

[22] http://www.w3.org/Submission/WSDL-S/

[23] http://www.wsmo.org/TR/d16/d16.1/v1.0/

[24] http://www.wsmo.org/TR/d13/d13.1/v0.3/

[25] http://wsmo4j.sourceforge.net/

[26] http://www.wsmostudio.org/

[27] http://sourceforge.net/projects/wsmx/

[28] http://sourceforge.net/projects/wsmt

[29] http://tools.sti-innsbruck.at/wsml2reasoner/

[30] http://tools.sti-innsbruck.at/wsml/dl-reasoner/v0.1/

[31] http://tools.sti-innsbruck.at/wsml/rule-reasoner/v0.1/

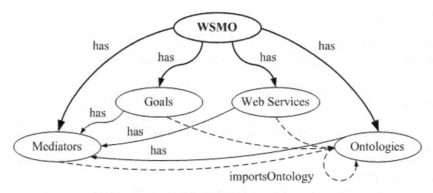

Fig. 2.6 Top elements of the WSMO ontology

Web services specifically define the non-functional, functional and behavioural properties of a web service. In WSMO, *non-functional property* is a broad concept, which may include a service's static information descriptions and quality properties[32]. The functionality of a web service is defined by its *capability*, which consists of *shared variables*, *pre-conditions*, *assumptions*, *post-conditions* and *effects* of the web service. The behaviour of a web service is defined by its *interface*. More details of web services will be elaborated with real examples in Section 2.4.1.

Goals consider the description of web services that would potentially satisfy users' desires, and were first proposed in WSMO. They are the representations of objectives for which fulfillment is sought through the execution of a web service. Similarly, goals have *capability* (showing what the user would like to have) and *interface* (meaning what the user would like to interact with). In Table 2.1, for example, the goal of booking a ticket is provided by an WSMO initiative[33], which has a text-based description saying "the goal of booking a ticket for a trip from Innsbruck to Venice". This goal has a capability with only one post-condition defining a reservation of a ticket purchased. The used terminologies, such as *Ticket*, *Trip* or *Reservation*, are defined by its imported ontologies.

Mediators are used to solve interoperability problems between goals, ontologies and web services, and are newly proposed in WSMO, too. WSMO distinguishes four different types of mediators: *ggMediators* (representing a refinement of a source goal into a target goal or state equivalence if both goals are substitutable), *ooMediators* (resolving possible representation mismatches between import ontologies), *wgMediators* (mediating a web service totally or partially fulfilling a goal) and *wwMediators* (mediating two linked web services).

To summarise, in WSMO the *ontologies* provide the basis of service semantics, the *web service* defines the capabilities of a service, the *goal* represents a user's desire of a service and the *mediators* resolve all possible mismatches during interoperation of web services. The ontologies and web service capabilities contribute to service

---

[32] http://www.developer.com/services/article.php/2027911
[33] http://www.w3.org/Submission/WSMO-primer/

Table 2.1 A travel goal in WSMO

---

namespace {_"http : //www.gsmo.org/dip/travel/goal#",

goal _"http : //example.org/havingATicketReservationInnsbruckVenice"

  nfp

    dc#description **hasValue** "booking a ticket from Innsbruck to Venice"

  endnfp

  importsOntology {tr _"http : //example.org/tripReservationOntology",

      loc _"http : //www.wsmo.org/ontologies/locationOntology"}

  capability

    postcondition

      definedBy

        ?*reservation* [*reservationHolder* **hasValue** ?*reservationHolder*,

             item **hasValue** ?*ticket*] **memberOf** *tr#reservation*

        and ?*ticket*[*trip* **hasValue** ?*trip*] **memberOf** *tr#ticket*

        and?*trip*[*origin* **hasValue** *loc#innsbruck*,

           destination **hasValue** *loc#venice*] **memberOf** *tr#trip*.

---

discovery, while service composition mainly considers the interfaces of web services and the ontologies used.

2.3 Semantic Annotations for WSDL and XML Schema

Semantic Annotations for WSDL and XML Schema (SAWSDL) is another popular way to provide semantics of web services. In [48, 82] a set of extension attributes for the Web Services Description Language (WSDL) and the XML Schema definition language are defined, that allow to describe additional semantics of WSDL components (e.g. interfaces, operations, faults) and WSDL type definitions (e.g. simple/complex types, element/attribute declarations). Using the notions of *model reference* and *schema mapping*, the SAWSDL specification indicates how to accomplish semantic annotation using references to semantic models, e.g. ontologies.

Taking an example from [43], in Fig. 2.7 the simplified code of the operation element's "order" annotation carries a reference to a concept (i.e. RequestPurchase-Order of Partner Interface Process (PIP)) in a semantic model (i.e. Rosetta purchase-order Ontology[34]), which provides a high-level description of the operation and specifies its behavioural aspects or includes other semantic definitions. SAWSDL provides an effective and practical way to annotate WSDL [28, 21].

2.4 Semantic Web Services and Ontologies

The notion **ontology** originates in philosophy. In artificial intelligence, an ontology is defined as *"an explicit formalisation of a shared understanding of a*

---

[34] http://lsdis.cs.uga.edu/projects/meteor-s/wsdl-s/ontologies/
rosetta.owl#

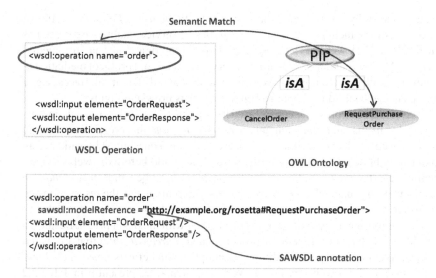

Fig. 2.7 Partial SAWSDL description of a service supporting the Purchase Order as defined in RosettaNet

conceptualisation" [60]. An ontology can represent a shared understanding of some domain of interest. Then it facilitates exchange and understanding of knowledge. Generally, an ontology has five components: domain *classes* (or *tasks*), concept *relations, functions, axioms* and *instances* as mentioned by [127].

The ontology concept is the cornerstone of the semantic web service technology. It does not only provide semantic description to web services, but by supporting logic it also provides capabilities for semantic expressiveness and reasoning, which enabling dynamic and automatic service discovery and composition.

A common characteristic of the emerging frameworks for semantic web services (e.g. OWL-S, WSMO, IRS) is that they combine two kinds of ontologies to describe web services. First, a *generic web service ontology*, i.e. the WSMO web service, defines generic web service terms (e.g. pre-condition, post-condition, assumptions, effects) and prescribes the backbone of the semantic web service descriptions. Secondly, a *domain ontology* specifies the domain knowledge of a kind of web services, especially defining all domain concepts, e.g. *Ticket*. We discuss these two kinds of ontologies in the next two subsections.

2.4.1 Generic Web Service Ontologies

Here we take WSMO web services as an instance to explain what a generic web service ontology looks alike. As defined by WSMO D2v1.3[35], WSMO web services have non-functional properties, capabilities and interfaces. Fig. 2.8 specifies the web

---
[35] http://www.wsmo.org/TR/d2/v1.3/

service elements by a UML class diagram, where the *non-functional* properties of web services may include cost-related properties or service qualities expressed as logical axioms[36]. A web service may import other ontologies and may use mediators, i.e. *ooMediator* and *wwMediator* (for service process and protocol mediations). Moreover, a web service might have one capability and different interfaces, e.g. a travel agent can be used for booking flights as well as browsing flights.

A web service has only one capability. The *capability* of a web service defines its functionality by the other four elements, viz. *pre-condition* (the information required before the web service's execution, e.g. the reservation request must be valid before booking a flight ticket), *assumption* (the state of the world before the web service's execution, e.g. the credit card provided for payment must be valid), *post-condition* (the information generated after the web service's execution, e.g. the reservation and an E-ticket will be confirmed) and *effect* (the state of the world after the web service's execution, e.g. the credit card will be charged with the ticket price). Besides, a capability might have non-functional properties and *shared variables* that are shared between pre-conditions, post-conditions, assumptions and effects. Instead of showing the *BookingTicket* web service[37] in WSML Schema, we present it in Table 2.2 in a more easily understandable way.

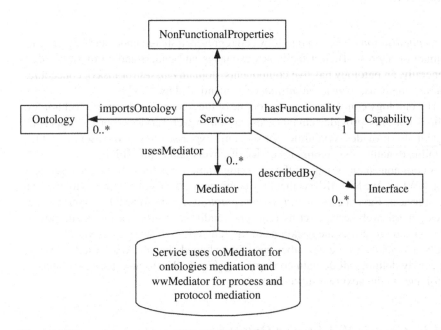

Fig. 2.8 Web service elements in WSMO ontology

[36] http://www.wsmo.org/TR/d16/d16.1/v0.21/20051005/
[37] http://www.w3.org/Submission/WSMO-primer/

Table 2.2 Capability of example web service

Service BookTicketService:
   ***Capability** BookTicketCapability:
          precondition: ReservationRequestIsValid,
          assumption: CreditCardIsValid,
          postcondition: (Reservation, Ticket)
          effect: PaymentIsConfirmed.
   ***Interface** ...
   ***WSDLGrounding** ...

The *interface* of a web service presents further information on how the service functionality will be achieved. It provides a twofold view on the operational competence of the web service: the *choreography* defines how to communicate with the web service in order to consume its functionality from the client's point of view, and the *orchestration* defines how the overall functionality is achieved by co-operation of more elementary web service providers.

Choreography and orchestration are defined in similar ways. Both consist of two parts, *state* and *guarded transition rules*. In our *BookTicketService* example, the *BookTicketChoreography* has three choreography transition rules (viz. c_tr1, c_tr2 and c_tr3 of Table 2.3), which specify their respective states by their *in* and *out* parameters. Taking c_tr1 as example, it only has a reservation request as *in*, and *out* is a temporary reservation for the user.

Similarly, orchestration bases on the case of the current input its decision for the next operation with the other services. Take c_tr2 as example, this orchestration transition rule defines that if the payment credit card is a golden one, then it will invoke *GoldenCreditCardService*.

Table 2.3 Interface of example web service

Service BookTicketService:
   *Capability ...
   ***Interface** BookTicketInterface
   choreography BookTicketChoreography
        $c\_tr1 : in(reservationRequest), Out(temporaryReservation)$
        $c\_tr2 : in(temporaryReservation, creditCardIsValid), Out(reservation)$
        $c\_tr3 : in(temporaryReservation, creditCardNotValid),$
                       $Out(negativeAcknowledgement)$
   orchestration BookTicketOrchestration
        $o\_tr1 : in(creditCardIsPlastic), Out(PlasticCreditCardService)$
        $o\_tr2 : in(creditCardIsGolden), Out(GoldenCreditCardService)$
   ***WSDLGrounding** ...

Grounding to WSDL. The details of WSMO grounding have been studied in the WSMO work draft D24.2v0.1[38]. Basically, elements of WSMO ontologies are

---

[38] http://www.wsmo.org/TR/d24/d24.2/v0.1/

bidirectionally mapped to XML elements using XML Schema, and the functional and behavioural service descriptions of WSMO are related to the description construct presented in WSDL (using built-in WSMO grounding properties or using the SAWSDL specification[39]). By such a grounding, WSMO web services can interoperate with currently deployed SOAP web services and client frameworks. For the example above, its WSMO choreography is grounded to its WSDL description. There are three WSDL operations generated via three choreography groundings, as listed in Table 2.4.

Table 2.4 Grounding of example web service

Service BookTicketService:
    *Capability ...
    *Interface ...
    *WSDLGrounding
        ChoreographyGrounding2wsdl : $Gr1$ $(c\_tr1 - > op1 : temporaryReservation)$
        ChoreographyGrounding2wsdl : $Gr2$ $(c\_tr2 - > op2 : reservation)$
        ChoreographyGrounding2wsdl : $Gr3$ $(c\_tr3 - > op3 : negativeAcknowledgement)$

2.4.2 Web Service Domain Ontologies

Externally defined knowledge plays a major rôle in any semantic web service description. As for WSMO web services, it offers a generic framework to describe web services, but to make them truly useful, a domain ontology is required.

The service ontology of our travel scenario is illustrated in Fig. 2.9. Concepts, such as *Airline Company*, *Ticket* and *Trip*, denote the main entities of the domain to be conceptualised. A set of relations between concepts should be specified. An important relation is the *isA* relation which indicates subsumption between two concepts. For instance, a *traveler* is a *person* who buys tickets. Several airline web sites can semantically be described with the concepts defined by this ontology. *AirNinja Inc.* as an instance of concept *Airline Company*, for example, describes a low-fare company which sells flight tickets from Innsbruck to Venice.

In WSMO, a web service domain ontology is formally defined by using a kind of WSML language. Moreover, a service ontology does not only define *concepts* and concept *relations* in order to capture semantic properties of relations and concepts, but it also provides a set of *axioms*, which are expressions in some logical language. Besides, *functions*, as special relations, allow to actually evaluate a function if concrete input values for the parameters are given. For instance, a part of the ontology imported by *BookTicketWebservice*, named *tripReservation* ontology, is listed in Table B.1 in Appendix B.

So far, we have briefly introduced generic web service ontologies and domain ontologies which are used for service representation.

---

[39] Semantic Annotations for WSDL, http://www.w3.org/2002/ws/sawsdl/

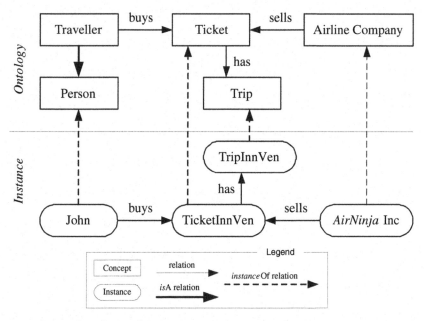

Fig. 2.9 An example service ontology and ontology instance

Chapter 3
Related Work

In this chapter we identify and present an overview of related work. Since the main goal of this book is to facilitate the discovery of semantic web services from the perspective of application ontologies, the work on semantic service discovery and, in particular, on ontology building and merging in the context of SWSs is relevant. Nevertheless, how to build common domain ontologies for semantic web services has not been addressed very well, although this issue is recognised by academia and industry since web service technology commenced to emerge.

Therefore, here we study the related work from two aspects, the methodologies to discover semantic web services, and the ontology engineering methods useful in the field of SWSs. In the course of the overview of previous efforts we state their contributions and problems, and incidentally we put forward our point of view in this regard.

3.1 What Is Web Service Discovery?

Although various proposals for discovering web services are available, the understanding of *discovery* is different, and has often been confused with terms such as *selection* and *matching*. Therefore, it is important for us to clarify first what precisely our understanding of web services discovery is.

In the literature, there are not many formal definitions of service discovery. One[1] states that *web service discovery is the process of finding a suitable web service for a given task*. Thus, discovering a web service is seen as a process. A more official definition of web service discovery (by W3C[2], 2004) reads

> *the act of locating a machine-processable description of a web service-related resource that may have been previously unknown and that meets certain functional criteria. It involves matching a set of functional and other criteria with a set*

---

[1] http://en.wikipedia.org/wiki/Web_Services_Discovery
[2] http://www.w3.org/TR/ws-gloss/

X. Wang and W.A. Halang: Discovery and Selection of Semantic Web Services, SCI 453, pp. 27–37.
springerlink.com   © Springer-Verlag Berlin Heidelberg 2013

of resource descriptions. The goal is to find an appropriate web service-related resource,

and a *discovered service* is a one that enables agents to retrieve resource descriptions related to web services. From the definition above, we see that web service discovery aims to find a service which meets a certain criterion required by a *goal* (or the service requirements provided by users).

In WSMO, researchers distinguish two terms, *web service* and *service* [52]. They argue that a *web service* is a computational entity able (by invocation) to achieve a user's goal, and in contrast a *service* is the actual value provided by this service invocation. Thus, there are both *web service discovery* (by considering the abstract and static characterisation of the kind of services that can be accessed via web services) and *service discovery* (based on the usage of web services for dynamically discovering actual services with concrete data). Hence, one distinguishes between *abstract and static* discovery and *dynamic or run-time* discovery.

For the terms *selection* and *matchmaking* an appropriate definition is provided by [19], where "*selection* is the choice of a matching service, possibly taking into account a ranking, the outcome of a negotiation and other run-time information; it may imply a number of subscriptions". Given a set of service descriptions and a query, "*matchmaking* is the activity that determines the set of service descriptions that fulfills all mandatory requirements expressed in the query".

From the above consideration, it is apparent that discovery, selection and matchmaking are different. As illustrated in Fig. 3.1, we define that semantic service discovery consists of a set of subactivities, viz. **location** of a number of service descriptions, **matchmaking** returning all matching service descriptions and **selection** of one or more services ranked among the matching ones.

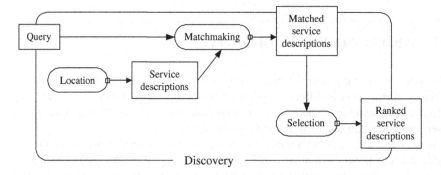

Fig. 3.1 Service discovery, matching and selection

In the following two sections, we introduce the work related to web service discovery, which adopts methods from component-based software engineering [183], information retrieval [167, 87, 57, 70], machine learning [122] and artificial intelligence [50]. The approaches considered include signature matching, schema matching, semantic matching and similarity assessment measures. In accordance with technological development, we divide the presentation into two stages, one of work related to the context of web services (Section 3.2) and the other one of work related to the context of semantic web services (Section 3.3) emphasising structural information and semantics, respectively.

3.2 Methods to Discover Web Services

The original intention of web service technology is to define web services in a machine-understandable way, in order to be automatically discoverable and invocable by other agents or services. Finding suitable web services is widely considered as a key task of web service technology, and its solution is important to service composition, invocation and execution. In the sequel, we outline the approaches used for web service discovery.

WSDL-related discovery. Since WSDL, as interface to access web services, supposes to provide the description of web services, the initial work naturally considered WSDL-based discovery. Based on XML Schema, WSDL documents define *service*, *operations* and operation *inputs* and *outputs*. Methods, such as text document matching, signature matching or schema matching, have been considered to measure the similarity of web services' WSDL descriptions.

Generally, *text document matching* is a method of matching documents based on weighted keyword similarity [74], term frequency analysis [180], text clustering [36, 178] or domain categories [12]. *Signature matching* is a means to retrieve and match functions (including function names, types and parameters) and modules of software components [182]. And *schema matching* is based on the semantics of schemas to find possible matches between two schemas, which normally includes linguistic and structural analysis [41, 42, 136].

For instance, [177] took a textual description of a desired service and used a structure matching method to identify the most similar services from a set of advertised ones. They actually utilised both the identifiers of WSDL descriptions and the structure of their operations, messages and data types to assess the similarity of two WSDL files. However, terms as the operation names of two zip-related web services, {*CalcDisTwoZipsKm, findZipCodeDistance*}, cannot be resolved by pure text analysis (e.g. with the string edit distance algorithm [145]) and structure matching, because without support of an ontology it is very difficult to measure the similarity of any two composed concepts, which prevalently appear in all descriptions of web services.

To summarise, the above methods are insufficient in the context of web services. For instance, text documentations for web service operations are highly compact, and ignore structural information that aids in capturing the underlying semantics of service operations. Consequently, discovery results obtained by text document matching are usually incorrect. Also searching for similar web service operations differs for signature and schema matching. First, the granularity of such search is different: operation matching can be compared to finding a similar schema, while schema matching looks for similar components in two given schemas assumed to be related. Secondly, the operations in a web service are typically much more loosely related with one another than are tables in a schema, and in isolation each web service has much less information than a schema.

UDDI-related discovery. Searching for web service registries is another intuitive way to discover web services. As the mainstream of registry technology, the UDDI[3] has defined two standard APIs for publishing and inquiring web services. The information web services publish to UDDI is defined by the *businessEntity* (information about the service publisher, e.g. names, URLs, short business description, contact, service categories), the *businessService* (descriptive information of services) and the *technical Models* (tModels, technical descriptions for web services provided and information of how to access them).

The inquiry API of UDDI provides a simple browsing-by-business-category mechanism to select web services. The searching mechanism mostly focuses on a single search criterion (i.e. business name, business location, business category), or on service type by name, business identifier or discovery URL. That is, UDDI is mainly based on category browsing and keyword-based search for businesses and services.

As a centralised searchable directory of web services, one disadvantage of UDDI is obvious. In order to avoid the bottleneck problem, many studies on distributed service discovery in peer-to-peer or grid environments were carried out, such as [143, 148, 151]. Measuring web service similarity, however, has always been the core issue, both in UDDI and in distributed environments, and apparently it was not solved by approaches related to UDDI.

Summary. WSDL- or UDDI-related web service discovery basically uses keyword-based and signature-based methods for web service similarity matching, which only consider the names of web services, but ignore their real functions and function semantics. It is essential, however, that service semantics are considered in web service discovery.

3.3   Methods to Discover Semantic Web Services

As already said, web services are discovered basically by measuring the similarity between service requirements (s_R) given by a service requester and service advertisements (s_A) from service providers. Hence, we define the similarity of web services as

$$simService(s_A, s_R) = match(s_A, s_R) \in [0, 1] \tag{3.1}$$

If $simService(s_A, s_R) > \tau$ (with τ being a threshold), then s_A is assumed to be a matching service.

---

[3] http://uddi.org/pubs/uddi_v3.htm

Although the technology to discover semantic web services should be grounded more on semantics, the current discovery mechanisms employ techniques similar to the one above, such as software component retrieval (signature matching and specification matching [32]) and information retrieval relying on textual descriptions of artifacts [47]. In the following, we coarsely divide the related work into several categories based on the preference of their similarity measurements.

1. *Capability-based approaches* are no longer limited to text and structure comparison, but consider the similarity of web service capabilities, including non-functional and functional properties. For example, a service is not considered acceptable if one finds an operation named *"bookFlight"*, only, but if such a web service is unable to perform the task *"booking a flight from Innsbruck to Venice on 24 Oct 2007 with a price less then 100€"* described in natural language. Such approaches usually represent service requirements and service advertisements in formal logic-based service description languages (e.g. DAML-S or OWL-S). Subsumption reasoning [20] is then used to measure their capability similarity.

   The foundation of this type of work was laid in [161], where discovery and similarity assessment methods for software components and agents were investigated, both by means of signature matching [134] and specification matching [32]. Their basic mechanism is that two components are considered to match if their signatures or specifications match. Specification matching considers the semantic information (or behaviour information) about a service. In their later work, the authors specified a language LARKS for defining both the agent advertisements and requests [162], and proposed a dynamic matchmaking among heterogeneous software agents. The matchmaking process includes five different filters (context matching, profile comparison, similarity matching, signature matching and constraint matching), and filters can freely be combined to result in θ-subsumption matching. The measurement results were further elaborated into four degrees of matching (i.e. *exact, plug in, subsumes* and *fail*) by [89, 124, 15]. In [124], similar methods were described, but based on the description language DAML-S (the predecessor of OWL-S). For these approaches there is, however, no guarantee that service specifications provided by programmers correctly and completely reflect their services' behaviour. Signature matching considers function types, only, but ignores behaviour, although sometimes two functions with the same signature can have completely opposite behaviour. Moreover, it is hard to motivate programmers to provide a formal specification for any component they write.

2. *Description Logic-based approaches.* Logical implication is the key to perform semantic matching in the context of semantic web services. Description logic (DL) [117] can be used to represent the concept definitions of an application domain (known as terminological knowledge) in a structured and formally well-understood way. Moreover, DL refers to logic-based semantics which can be translated into first-order predicate logic. Therefore, DL has become a cornerstone of the semantic web for its use in the design of ontologies. In fact, OWL-DL and WSML-DL are two DL-based languages.

In [65, 77, 89], logic expressions and reasoning about service descriptions are advocated for, using logical service description languages (e.g. DAML-S, OWL-S or WSML) and their respective reasoners to compare service descriptions. To represent the semantics of service descriptions by the DAML-S ontology, for instance, [89] used the DL reasoner Racer [64] to assess ontology-based service descriptions. Similarly, the approach proposed by [77] used WSML together with the *F-Logic* [78] reasoning engine *Flora-2*. To summarise, these approaches formulate service profiles in logic formats, and reason about their similarity based on subsumption trees given by reasoners (e.g. Racer or FaCT [69]).

Still, many problems remain with this type of semantic service discovery. Just to mention a smaller one, DAML-S service profiles contain too much information for effective matching [89], and the bigger problems are related to the logics of checking concept instances and instance relations (e.g. whether a relation holds between two instances). In conclusion, we argue that the result of logic-based matchmaking fails to suggest the web service matching best.

3. *SPARQL-based approaches.* From another point of view, semantic web services description languages actually use RDF-based triples to represent information. Therefore, employing SPARQL could be a way to query and access services [72, 150, 54, 172]. Adding a preprocessing stage is proposed in [54]. This is based on two SPARQL queries which, in order to improve the overall performance of service discovery, filter service repositories, in doing so discarding service descriptions that do not refer to any functionality or non-functional aspect requested by the user before the actual discovery takes place. While [150] uses SPARQL as a formal language to describe the pre-conditions and post-conditions of services, to optimise service discovery they also show that the evaluation of SPARQL queries can be used to check the truth of pre-conditions in a given context, they construct the post-conditions that will result from the execution of a service in a context and determine whether a service execution with those results will satisfy the goal of an agent. This book's primary concern is that all similar work actually bases more or less on the assumption that all services share the same set of ontologies.

4. *QoS-based approaches.* Qualities are very complex factors of semantic web services. When desired service capabilities are fulfilled, the service qualities might be the determining factors on whether to invoke a service or not. In [105, 141, 171, 185, 63, 160, 139, 184, 101] it was particularly stressed to consider service qualities in the course of service selection.

By [105], for instance, the objective QoS attributes (e.g. reliability, availability and request-to-response time) and the subjective ones (focusing on user experience) were enumerated in detail, and an agent framework coupled with a QoS ontology to address the selection problem was presented. Work of [88] and [141] emphasised the definition of QoS properties and metrics. By [141] all possible quality requirements were discussed and classified into several categories, including the ones related to run-time, transaction support, configuration management, cost and security. Also, some definitions and metrics of these quality properties were provided very briefly.

Although the QoS properties of SWSs were examined thoroughly, QoS ontologies developed and various QoS metrics and their measurements with respect to semantic services identified, unfortunately, quantifiable measurement techniques for semantic service selection have not been derived.

5. *SAWSDL-based approaches.* SAWSDL constitutes a practical way to add semantic annotations to web service descriptions. Some authors have, therefore, considered to use SAWSDL to discover web services [173, 80, 152, 55]. In [152] the reasoning-based SAWSDL matchmaker LOG4SWS.KOM is presented as an approach to map web service matches with discrete degrees of matching (as obtained by, e.g., subsume or plug-in) to a continuous scale (e.g. a real number in [0, 1]). As a hybrid semantic matchmaker for SAWSDL, SAWSDL-MX2 [80] considers three kinds of service similarities, viz. logical (applying conventional logical matching filters, which are exact, subsume, subsumed-by, plug-in and fail), textual (focusing on semantic annotation of the signatures) and structural (pertaining to the web services' XML structure).

Obviously, their common problems are that SAWSDL lacks a uniform, formal ontology language, either OWL-S or WSML, and that SAWSDL allows multiple references to different kinds of ontologies for annotating even the same service description element. The lack of uniform application domain ontologies renders the discovery of semantic service very difficult, because ontologies constitute an issue of central importance for the application of semantic web services.

6. *Ontology-based approaches.* The semantics of web services are implicitly contained in the ontologies used by service descriptions. For example, [44] and [67, 138] considered service similarity from the point of view of ontologies and ontological concept similarity. In [44], the similarity of concept terms is measured by a clustering algorithm according to their association degrees expressed in terms of their conditional probabilities of occurrence. In [67], the similarity of web services is measured by taking into consideration all aspects of OWL object constructors present in the format of the Resource Description Framework.

This type of work is very important and has touched the core of the problem of semantic service discovery, viz. semantic web services are using heterogeneous ontologies even for the same kind of services. Moreover, the methods of defining service ontologies for web services are being transferred to the service application developers. Hence, misunderstanding among semantic services is inevitable and obvious.

The previous three approaches avoid this problem by having a latent and fatal hypothesis, viz. advertised and required services share a single ontology. This hypothesis is, however, not true, and it is a great challenge for semantic web service technologies to come up with better solutions.

Unfortunately, the current ontology-based approaches are weak. Basically, they consist of finding ontological concept mappings between two pieces of service ontologies to further measure service similarity. We argue that a small piece of a service ontology contains very limited information (which is defined just for the purpose of service description), and that appropriate domain knowledge is not explicitly provided for semantic service discovery, yet. Without support by a rich domain ontology, ontology mapping of two pieces of service ontologies is unable to solve the heterogeneous problems of semantic web services and to contribute to service discovery.

Summary. In most cases, the methods outlined above are employed in combination [63, 160, 137, 75]. Any of the above approaches has specific limitations which, in turn, revealed the aspects which should be considered in course of the future development of semantic web service discovery, viz. capability, quality and ontologies. To make up for these deficiencies, in Chapter 4 we shall propose a semantic service model apt for selection purposes, which specifically defines the capability and quality factors of services, and we shall present matchmaking algorithms for this service model aiming to solve the problem of quantitatively measuring semantic service similarity, which was left unsolved by the DL-based approaches.

Building a uniform application ontology for a kind of web services is our main contribution to semantic web service technology to be described in Chapter 5, and in Chapter 6 we shall further present our approach to ontology-based service discovery based on comprehensive application ontologies built to resolve the mismatch of heterogeneous service ontologies, which is the issue raised by the other related ontology-based approaches. As service qualities have complicated features and there is no uniform measuring standard, in Chapter 7 we shall present a normalisation algorithm to evaluate multiple quality metrics in combination, which is still missing in the QoS-based approaches.

3.4   Methodology to Build Ontologies

Automatically building application domain ontologies for a kind of semantic web services is the most important part of our work, because it constitutes the foundation for improving semantic service discovery. The approach taken to build application ontologies is actually to merge individual service ontologies. Therefore, the related work to be considered should include ontology building and learning technology from traditional areas such as ontology engineering and artificial intelligence.

3.4.1   Ontology Learning Methods and Tools

Building an ontology through learning is the most traditional approach. In ontology engineering, ontology learning means to acquire knowledge from experts of a given domain or through some kind of (semi)automatic process [58]. Most of the existing approaches try to learn domain ontologies from *textual sources* (e.g. text or corpora [26, 153], from the web [146] and web sites [179]), from *instances* (particular instances taken from files) [113] or from *schemata* (e.g. relational database schemas, entity-relationship models and XML schemata, since they can be used to generate ontologies by a re-engineering process). These approaches are normally carried out by means of natural language analysis and machine-learning techniques.

For learning ontologies from text corpora [97], five methods were developed, namely *pattern-based extraction* (concept relations are recognised in sequences of words following given patterns) [68], *association rules* (used to discover non-taxonomic relations between concepts using statistics of term concurrences in texts) [96], *concept clustering* (concepts are grouped according to the semantic distance between terms) [110], *ontology pruning* (eliciting a domain ontology by using a core ontology and a corpus, or several text corpora) [76] and *concept learning* (a given taxonomy is incrementally updated as new concepts are acquired from real-world texts) [66].

Accordingly, a number of ontology-learning or text-mining tools was developed to construct ontologies, such as *Mo'k Workbench* [18], *OntoLT* [26], *Text2Onto* [34], *ASIUMsystem* [49], *TextToOnto* [100] and *OntoLearn* [118]. These tools were compared by [34] and [146]. Here, we omit a detailed analysis and summarise these tools for our proposes as follows:

- Many of these tools construct ontologies from textual resources and depend on techniques of natural language processing to identify concepts.
- Most of them combine machine-learning approaches with basic linguistic processing, such as tokenisation, lemmatising and shallow parsing.
- Although depending on different paradigms, such as clustering [18, 49], probabilistic ontology modeling [34], native Bayesian theory in connection with support vector machines [123] or natural language processing [146], the learning process similarly includes term extraction from text, ontology aligning and ontology pruning.

In the context of SWSs, however, service ontologies are generally defined in schema-based and well-structured documents in formal languages, e.g. OWL or WSML. Significant terms are already identified. Normally, the only text processing needed is breaking strings at capitalisation or non-alphabetic characters. But we have seen no real ontology learning tool designed for semantic web services.

3.4.2 Ontology Merging

For SWSs, service providers normally attach small pieces of service ontologies when publishing their web services. Thus, there are multiple service ontologies and services even for the same kind of application domain. This makes it possible to build an application ontology by merging existing service ontologies and, as we know, they can be well written with web ontology languages instead of text. Therefore, the challenge of building an application of SWSs is actually transformed into the problem of merging ontologies.

An early approach to support the merging of ontologies was published in 1998 [71]. Several heuristics were described there to identify corresponding concepts in different ontologies, e.g. comparing the names of two concepts, comparing the natural language definitions of two concepts by linguistic techniques and checking the closeness of two concepts in a concept hierarchy [98]. More recently, other approaches relying on syntactical and semantic matching heuristics were proposed, including OntoMorph [30], Chimaera [106], PROMPT [120], ONIONS [156] and FCA-merge [159]. Taking PROMPT and FCA-merge as examples, we explain the current situation in the field of ontology merging.

PROMPT is a semi-automated method of ontology merging and alignment embedded in the Protégé-2000 tool developed by the Stanford Medical Informatics Group in the year 2000. Under continuous user interaction in selecting suggested operations, it merges ontologies (including merging classes, meta-classes, slots, bindings between a slot and a class or deep/shallow copying a class from one ontology), incorporates changes and finds conflicts of the resulting ontologies.

FCA-merge was developed by Institut für Angewandte Informatik und Formale Beschreibungsverfahren of Universität Karlsruhe in 2001. It operates by entering a mass of instances of two ontologies to be merged, and uses methods of formal concept analysis (FCA) for the proper merging. It currently works for light-weight ontologies, only.

The shortcomings of the current approaches to ontology merging are: (1) lack of a standard for merging and dealing with conflicts, (2) being only semi-automatic, as human interference is needed, (3) only working on light-weight ontologies without sufficiently considering concept axioms and constraints and (4) limited suitability for further usage, such as producing domain knowledge bases providing powerful reasoning capabilities.

3.5   Discovery Frameworks for Semantic Web Services

So far, many discovery architectures for semantic web services have been built with their respective discovery mechanisms. In this section, we briefly introduce several typical discovery systems.

OWLS-MX. This hybrid matchmaker was described in [79], and may be thought of as an implementation of an OWL-S-based service discovery framework. It represents an approach to hybrid matching of semantic web services that complements logic-based reasoning on respectively annotated content and services with approximate matching by similarity computations based on syntactic information retrieval.

The OWLS-MX matchmaker matches input and output to OWL-S services exploiting parameter values of *hasInput* and *hasOutput*, only. The authors evaluated their work by carrying out two comparative experiments. First, they treated the meaning of concept expressions in service descriptions as a bag of words, which can be measured by using the relative frequencies of indexed terms in expressions, and exploited string-edit or token-based similarity metrics from the information retrieval area with associated term-weighting schemes. Then, they fully considered their logical expressions and applied five degree filters in service matchmaking. Their results gained in comparative experiments provide some evidence in favour of the proposition that building matchmakers for semantic web services on Description Logic reasoners can improve the efficiency of service matchmaking.

The OWLS-MX matchmaker is implemented in variants of Java, using OWL-S 1.0 and the tableaux OWL-DL reasoner *Pellet* developed at the University of Maryland[4]. It is available as open source software under[5].

There does not exist, however, an implementation of a matchmaker that performs an integrated service *IOPE* matching by means of additional reasoning on logically defined pre-conditions and effects. Related work on logic-based rule languages for the semantic web, such as *SWRL* and *RuleML*, is under way.

METEOR-S. Built by the Large Scale Distributed Information Systems Laboratory at the University of Georgia, METEOR-S is a framework for semi-automatically marking up web service descriptions with ontologies [121]. With respect to discovery, METEOR-S addresses the problem of discovering services in a scenario where service providers and requesters may annotate their WSDL files with different ontologies. This approach relies on annotating service registries for a particular domain, and exploiting such annotations during discovery.

WSPDS. The WSPDS system [8] features a peer-to-peer discovery mechanism with semantic-level matching capability. This framework is guided by the principle that a decentralised design for web service discovery is more scalable, fault-tolerant and efficient than a centralised approach (e.g. UDDI). Also, WSPDS semantically annotates WSDL files using the WSDL-S framework described in [4]. One advantage of this approach is that it makes WSDL-S files agnostic to any ontology representation language (e.g. OWL-S or WSMO). On the other hand, adopting such a framework means that WSDL files would have to be re-written for the existing web services constituting additional overhead.

WSMX. The Web Services Execution Environment[6] is claimed to support dynamic mediation, selection and invocation of web services. WSMX is the reference implementation of the Web Service Modeling Ontology (WSMO). The current version of WSMX includes a rather basic means for discovery. Part of this functionality is provided by the WSMX Matchmaker. The previous version of this component implemented a very simple matchmaking algorithm based on string matching. An integration of a discovery and selection component into WSMX was developed and will be presented in Section 4.3.

Summary. As mentioned in Section 2.2, WSMX is chosen as implementation and evaluation environment for this work. From Chapter 4 on, we shall elaborate step by step our approach to ontology-based discovery of semantic web services in the context of WSMO and WSMX. In the course of this, we shall take full account of all general principles, such that our semantic service model and ontology-based matchmaking methods will also be suitable for the other semantic web service frameworks.

---

[4] http://www.mindswap.org

[5] http://projects.semwebcentral.org/projects/owls-mx/

[6] http://www.wsmo.org/

Chapter 4
A Service Model Facilitating to Discover Semantic Web Services

The first contribution of this book on defining a semantic *service model* (SM) is to be presented in this chapter, which aims to support discovery and selection of semantic web services. Throughout the chapter, the questions why a service model is needed, how it looks alike, how services are discovered based on such an SM and how it can be implemented will be answered.

4.1 Conceptual Framework for Service Discovery

Four rôles need to be specified for the discovery of semantic services, namely service providers (abbr. *SPs*), service requesters (abbr. *SRs*), a service search engine (abbr. *SE*) and a domain ontology. The *SPs* are supposed to publish web service descriptions as service advertisements s_A, and an *SR* sends its service request, which is formalised into a *goal* as s_R, to an *SE*. The *SE* is responsible for matchmaking between web services and goal. Besides, a *domain ontology* is the application knowledge shared by web services and goal.

Scope of this approach. How to obtain and represent user service requirements as formal goals in a specific service description language is quite complex a topic outside of our scope. Here we assume that user requirements have already been described as a *goal* in the same language as the web services, e.g. OWL or WSML. On defining goals, some interesting ideas on representing user requirements (normally given in natural language) as abstract goals are proposed by [13]. For finding a real goal instance based on an abstract goal we refer to the doctoral thesis [157]. But essentially, goal-service similarity is still a problem as is service-service similarity, which is addressed in this book.

As discussed in Section 3.1, web service discovery is not a combination of simple activities such as locating services and matchmaking, but actually it has very complex subprocesses. As WSMO [24], we also distinguish between *web service*

X. Wang and W.A. Halang: Discovery and Selection of Semantic Web Services, SCI 453, pp. 39–61.
springerlink.com   © Springer-Verlag Berlin Heidelberg 2013

discovery and *service selection*. As a computational entity, for us a *web service* is an abstract description of service capabilities in a logical language; and a *service*, in contrast, is an instance with its actual values provided upon service invocation.

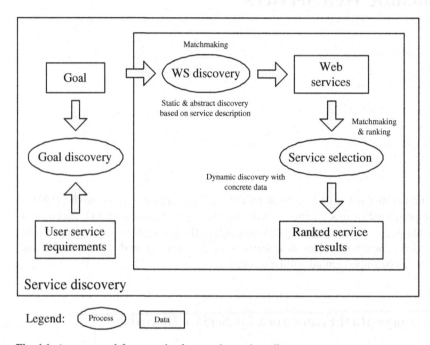

Fig. 4.1 A conceptual framework of semantic services discovery

An integrated conceptual framework to discover semantic web services is proposed in Fig. 4.1, in which the data of a discovery process are *user service requirements*, *goal*, *web services* and *services*. The subprocesses include *goal discovery*, *web service discovery* and *service discovery*. This book will only focus on the content inside the dashed box.

4.1.1 Two-Phase Discovery

During run time, there are two phases of finding all matching services.

Phase I. Static *web services* discovery at an abstract description level. Here, web services are matched at an abstract level taking into account the capability descriptions of web services and goal. The result will be a set of candidate domain web services (as *Resultset*₁), which have all the potential to fulfill the given service requirements. This phase mainly benefits from *ontology mediation* and *reasoning*. The ontology mediation is considered from the data level when different ontologies are used in service descriptions. As web service descriptions are defined in a certain logical language, e.g. Description Logic, several set-theoretical relationships will be used here to measure the similarity of services and goal, which are graded as exact match, plug-in match, subsumption match, intersection match or disjointness (cp. Section 4.3.2).

Phase II. Dynamic *service* selection at instance level.   This stage is based on the usage of web services to discover actual services. That is, with the real data from the user the candidate services of *Resultset*$_1$ are queried dynamically and, further, the service matching best is selected. This phase mainly benefits from quality-based service selection algorithms measuring the similarity of services and goal and ranking the final service set, noted as *Resultset*$_2$.

Obviously, there are many factors contributing to semantic service discovery, and different factors have different effects on these two discovery phases. In order to clarify the factors involved in semantic service discovery, in the following section we define a service model apt for service discovery and selection.

4.1.2 Assumptions

Before going into the details of the subject, it is necessary to declare some assumptions on which our semantic service model is based. We assume the following.

1. Service requirements have already been customised into a *goal* in the same format and language as web services.
2. *Web service* and *goal* share the same domain ontology to define their capabilities. Since web services are, in fact, using different service ontologies, in Chapters 5 and 6 we solve the problem of heterogeneous ontologies. Hence, for service discovery we can assume that a single application domain ontology is used.
3. Service discovery is normally an environment-independent task, i.e. it may occur in peer-to-peer, grid, ad hoc networks and so on. In order to easily explain our work, we assume that web services are centrally published in a service registry, which acts as a web service repository.

4.2 Service Model for Discovery

Without regard to any concrete implementation system and description language for defining a semantic service, its *information semantics, non-functional properties* and *functional properties* should be described. The information semantics refer to the service ontology, which defines an agreed common terminology for web services by providing concepts and concept relationships (cp. Chapter 5). Normally, the static information (like *service name, service categories* or *a short textual description*) is considered as the non-functional properties of web services; while the information about its operations is defined as its functional properties, which include *inputs* and *outputs, pre-conditions* and *effects* (IOPEs). Besides, there is some information on service quality features, such as security, integrity and reliability. Although some work, e.g. WSMO, classifies them as non-functional features, for the reason of selecting services we separate them as a set of independent features.

Example. A flight-booking web service is denoted by a service name, sometimes with a short service description, and advertises its ticket browsing and booking functions. If a user wants to invoke this service, he/she must provide some information required by the service as inputs (e.g. origin and destination, date of flight, number of passengers and so on). The pre-condition is that the user has to have a valid credit card for payment. The final effects of the service's invocation are that the cost is drawn from the user's credit card, and that an electronic flight ticket is issued. The reputation and security of the service could be its most important qualities.

Based on the above analysis, the service model is defined as a tuple,

$$s = (NF, F, Q, C) \tag{4.1}$$

where NF is the set of the non-functional properties of a web service, F is the set of the service's functional properties, Q defines its quality features and C is the possible cost of invoking it. As cost C, one of the quality features, often becomes the ultimate determinant factor for selecting a web service, our service model separates it from Q and explicitly stresses it in order to improve the selection efficiency of semantic web services.

In the following section, each element of this service model will be defined in detail and illustrated with a running example. Since both OWL and WSML are partly rooted in Description Logic (DL) [7], without loss generality the syntax and semantics of DL is basically used by this service model.

4.2.1 Definitions

Let $\Sigma = \{0, 1, ..., 9, a, b, ..., z, A, B, ..., Z, \_, -\}$ be an alphabet, a non-empty finite set[1]. A *string* (or **word**) over Σ is any finite sequence of characters from Σ. For instance, "travelAgency" is a string over Σ. The set of all strings over Σ of any length ($n \in \mathbb{N}$) is the Kleene closure of Σ, denoted by Σ^*. In terms of Σ^n,

$$\Sigma^* = \bigcup_{n \in \mathbb{N}} \Sigma^n \tag{4.2}$$

For the sake of simplicity, we denote the string set Σ^* by L. Therefore, in the context of semantic web services, a **service name** (sn) is a string identifer denoted as $sn \in SN$, $SN \subset \mathscr{P}(L)$; a **service category** (sc) is a string identifer denoted as $sc \in SC$, $SC \subset \mathscr{P}(L)$; and a **service description** (sd) in form of a short **text** readable by humans is a collection of strings denoted as $sd \in SD$, $SD \subset \mathscr{P}(L)$.

---

[1] http://en.wikipedia.org/wiki/String_(computer_science)

Let *DT* be the set of data types. Except of the common data types, e.g. *String, Integer* and *Date*, *DT* also allows any other defined data types by combining multiple elements of other types. For example, a new data type named "Person" specifies that data interpreted as a person would include a name and a date of birth. Variables of a service function are identifiers that point to values, and all variables have associated data types. We denote a variable and its data type as $vr : dt$.

Let *ass* be a DL assertion. We use DL assertions on concepts (called TBox, e.g. pre_1 of Expression (4.4)) and individuals (called ABox, e.g. pre_2 of Expression (4.5)) to formally represent the knowledge base of web services. For the syntax and semantics of DL *TBox* and *ABox* assertions refer to Appendix A.

Let *Ass* be a set of DL assertions and $ass_x \in Ass$, $x \in N$, and let ρ be a predicate symbol of the set $\{\vee, \wedge, \neg\}$, then a logical expression *le* (or formula) is defined as

$$le = ass_x \mid \rho(ass_1, ass_2, ..., ass_x)^* \quad (4.3)$$

Logical expressions are used to express service pre-conditions (e.g. Expression (4.6)) and effects. For example, booking on-line a discount flight for a child has two requirements, viz. the age of the user to be $2 \sim 12$ and that he/she is a European citizen. They can be written as follows,

$$pre_1 : (\geqslant 2hasAge.Human) \sqcap (\leqslant 12hasAge.Human) \sqsubseteq Child \quad (4.4)$$

$$pre_2 : isEUCitizen(Human) \quad (4.5)$$

$$pre_1 \wedge pre_2 : (\geqslant 2hasAge.Human \sqcap \leqslant 12hasAge.Human) \wedge isEUCitizen(Human) \quad (4.6)$$

where $\geqslant$ and $\leqslant$ denote the minCard (at-least restriction) and maxCard (at-most restriction) number restrictions, respectively, as defined in DL (cp. Appendix A).

4.2.1.1 Non-functional Property *NF*

For the propose of effective service selection, we abandon the complexity of non-functional service properties and take just three factors into account. Then, the *NF* of Definition 4.1 is succinctly defined as a triple,

$$nf = < sn, sc, sd > \quad (4.7)$$

and $nf \in NF$. Definition 4.7 reveals the fact that a web service is allowed to be published with more than one service name and category, which depends on what the service provides. Obviously, a web service may also have more than one service description. Besides, the service category must be any possible category which is classified according to its application domain, be calculated on the basis of web service annotations as in [11], or be determined according to some standard industry taxonomies such as the one of the *North American Cartographic Information*

Society (NACIS[2]) or the *United Nations' Standard Products and Services Code* (UNSPSC[3]).

4.2.1.2 Functional Property *F*

A web service normally provides a set of functions as $F = \{f_1, f_2, ..., f_i\}, i \in N$. As discussed in Section 4.2, the function f_i is a quintuple

$$f_i = < Op_i, I_i, O_i, P_i, E_i > \tag{4.8}$$

with:

1. Op_i is a string and identifies the service operation i.
2. I_i is a set of variable identifers wrapped with data types, consisting of all inputs of Op_i:

$$I_i = < Ivr_{i1} : dt_1, Ivr_{i2} : dt_2, ..., Ivr_{ij} : dt_j >, dt_j \in DT, j \in N, \tag{4.9}$$

3. O_i is a set of variable identifers wrapped with data types, consisting of all outputs of Op_i:

$$O_i = < Ovr_{i1} : dt_1, Ovr_{i2} : dt_2, ..., Ovr_{ik} : dt_k >, dt_k \in DT, k \in N, \tag{4.10}$$

4. P_i is a logical expression of the pre-conditions of Op_i, named le_{prei}.
5. E_i is a logical expression of the effects of Op_i, named le_{effi};

4.2.1.3 Service Qualities *Q*

A set of properties is used to describe service qualities. Let q_t be the quality t of a web service, and

$$\mathscr{Q} = \{q_1, q_2, ..., q_t\}, t \in N. \tag{4.11}$$

Let **Q** be the conjunction of ABox assertions on q expressed in DL, named le_{qos}.

Almost all QoS properties of web services were identified by [88], such as *performance, reliability, scalability, capacity, robustness, exception handling, accuracy, integrity, accessibility, availability, interoperability, security* and *network-related QoS*. For system optimisation, we divide them into two categories from the customer's perspective, which are a necessary category $\mathscr{Q}_n$ and an optional one $\mathscr{Q}_o$, $\mathscr{Q} = \mathscr{Q}_n \cup \mathscr{Q}_o$.

- As a subset of $\mathscr{Q}$, $\mathscr{Q}_n$ only consists of a small number of service qualities, which are considered as the common and necessary ones for service selection and

---

[2] http://www.nacis.org
[3] http://www.unspsc.org

invocation. Moreover, a default $\mathcal{Q}_n$ is defined as $\mathcal{Q}_d$, having only the six features $\{qualityResponseTime, qualityExecutionTime, qualityReliability, qualityException Handling, qualityAccuracy, qualitySecurity\}$.

- As the complement of $\mathcal{Q}_n$, $\mathcal{Q}_o = \mathcal{Q} \setminus \mathcal{Q}_n$ consists of all remaining quality properties called optional ones.

Intuitively, the necessary qualities comprise the features typically required by customers. For instance, a customer who wants to find an on-line flight booking service is more concerned about the response time to his/her order rather than about the scalability of the service itself (of which the service developers should take care). Therefore, the response time is emphasised in the user service requirements, and is defined as a necessary one.

The necessary and optional QoS categories are changeable and customisable. If an end user stressed in his/her service requirements some specific qualities, which are not belonging to the default settings of necessary attributes, these qualities are moved from the category optional to the category necessary during service selection. This kind of design renders high extensibility and flexibility. In practice, our service model actually uses $\mathcal{Q}_n$ replacing $\mathcal{Q}$ to improve service selection efficiency, and this is why we differentiate between $\mathcal{Q}_n$ and $\mathcal{Q}_o$.

4.2.1.4   Service Cost C

The quantity C is a pre-assessed overall cost of consuming a web service, defined as a DL assertion as well and named le_{cost}. The cost of a web service normally depends on the specific service policies and application domains. Without loss of generality, the service cost is roughly calculated as,

$$C_{service} = C_{product} + C_{tax} + C_{handingFee} \tag{4.12}$$

Moreover, a web service's cost is often related to its quality. Faster, reliable, secure services will be more expensive. Besides, there could also be penalty cost associated with not meeting certain QoS goals or service level agreements (SLAs) based on service policies.

Actually, the cost C is defined as a replaceable, scalable element in our service model, aiming to optimise or quicken the service selection process. It can be replaced by any other element considered important to service selection. As the price (or cost) is normally the ultimate factor for invoking a web service, we define it explicitly in our model. That is, if the cost of a service far exceeds the user's quote, then this service will not be further considered by our service engine, and will be filtered out right away.

After having defined the service model, now a travel scenario is applied to explain our approach in the following section.

4.2.1.5   A Travel Scenario

Professor John Smith is planning to attend an international conference in Orlando, Florida, USA, during 14–17 July 2007. John lives in London, UK. Presently, he has to spend quite a time manually arranging his trip via a web browser, including searching for on-line booking services, selecting and comparing them, then deciding to take one to book his travel. With the constraints on venue and time of the conference, John further needs to reserve a room in a nearby hotel and, perhaps, John would also like to hire a car locally.

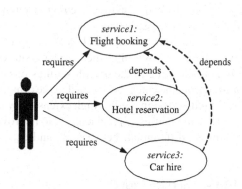

Fig. 4.2 A travel scenario

Such a travel scenario is actually quite a complex service application, because not only service discovery is involved, but also the composition of *service1:Flight Booking*, *service2:Hotel Reservation* and *service3:Car Hire* (cp. Fig. 4.2) should be considered. For example, *service3* is based on the destination of *service1* to discover the best car hire service, *service2* depends on the conference location to reserve a hotel room and, in a workflow, *service2* and *service3* can be parallel. The study of how to compose these three services is outside the scope of this book. But, as the discovery of each web service is the basis for service composition, our method can greatly contribute to web service composition. Therefore, we only take *service1* as an example to explain the process of service discovery.

4.2.1.6   Running Example: Booking a Flight

Strongly supported by industry, there are many flight web services which can meet John's requirements with similar functionalities. These services are normally developed and deployed using specification standards such as WSDL and OWL-S or WSML.

Goal example. By applying our service model for discovery, John's requirements are defined as a goal, which is:

- **NF** $= \{< \textit{"FlightBooking","Travel"}, \textit{FlightBookingDescription} >\}$. Note that *FlightBookingDescription* will be a short text about the anticipated service, e.g. "book the cheapest flight".
- Obviously, the required service function is $\mathbf{F} = \{f_1\}$, where F has only one operation as described in Table 4.1. The details of John's requirements include that the total price of a return flight should not exceed 800€ (including the ticket price), that the departure time should be between 8:00 and 20:00 hrs., that it is for one adult, no children and no infants, and that an economy class ticket is required.
- About the qualities of service $\mathbf{Q}$ we assume that John cares about the service response time, which should not be too long (e.g. being less than 15 sec), its reputation (e.g. its reputation rank is better than rank 6) and exception handing, and that it is safe. Then, $le_{qos} \doteq \cap (\leqslant 15 second.qualityResponseTime) \cap (\geqslant 6rank.Reliability) \cap (\geqslant 6rank.ExceptionHandling) \cap (\geqslant 6rank.Security) \cap (\geqslant 8rank.Reputation)$
- The service cost $\mathbf{C}$ should not exceed 800€, i.e. $le_{cost} \doteq (\leqslant 800cost.Service)$. Normally, we assume the price given by the user as maximum of the service cost.

Table 4.1 A goal defining John's travel requirements

$Op_1 \doteq < FlightBooking >$
$I_1 \doteq < (LondonUK, String), (OrlandoUSA, String),$
$\quad (July1407, Date), (July1707, Date), (1, Integer), (0, Integer),$
$\quad (0, Integer), (Return, String) >$
$O_1 \doteq < (Ticket, String) >$
$le_{pre1} \doteq \cap (\leqslant 800cost.Ticket) \cap (\geqslant 8departTime.Ticket)$
$\quad \cap (\leqslant 20departTime.Ticket) \cap (\geqslant 2ndClass.Ticket)$
$\quad \cap (isValid(CreditCard))$
$le_{eff1} \doteq \varnothing$
$le_{cost} \doteq (\leqslant 800.costService)$

Web service examples. Accordingly, Tables 4.2[4] and 4.3[5] present two real flight booking services s_1 and s_2 as examples, which have been re-described using our service model defined in Definition 4.1. For example, service s_1 provides an operation named *cheapFlightSearch* with a set of inputs and one output parameter, one pre-condition and one effect.

From these two service descriptions, we know that service s_1 will return an instance of "ticket" as result (which is defined by its service ontology, say so_1, cp. Table 4.4) and that service s_2 returns an instance of "travel voucher" (defined by its service ontology, say so_2, cp. Table 4.5). Although different service ontologies are used, they are describing very similar information. For instance, the concept *Cost* of so_1 is actually equivalent to the concept *Amount* of so_2. Such questions of data mediation between service ontologies will be elaborated in Chapter 5.

---

[4] *Serviceadvertisement*₁, at www.squareroutetravel.com/Flights/cheap-flight-ticket.htm

[5] *Serviceadvertisement*₂ at www.travelbag.co.uk/flights/index.html

Table 4.2 Service advertisement s_{A_1}

$NF \doteq \{< CheapFlightSearch, Travel, "Searching \ for \ cheap \ overseas \ flights." >\}$
$Op_1 \doteq< cheapFlightSearch >$
$I_1 \doteq< (LondonUK, String), (OrlandoUSA, String), (14 - 07 - 07, Date),$
$\quad (17 - 07 - 07, Date), (1, Integer), (0, Integer), (0, Integer),$
$\quad (ReturnTrip, String), (TravelInsuranceID, String) >$
$O_1 \doteq< (TravelVoucher, String) >$
$le_{pre1} \doteq (isValid(PaymentCard))$
$le_{eff1} \doteq (\geqslant 0(amount.CreditCard - cost.Ticket))$
$le_{qos} \doteq (\leqslant 8second.ResponseTime) \cap (\leqslant 8second.ExecuteTime)$
$\quad \cap (\geqslant 8rank.Reliability) \cap (\geqslant 7rank.ExceptionHandling)$
$\quad \cap (\geqslant 9rank.Accuracy) \cap (\geqslant 8rank.Security)$
$le_{cost} \doteq (= 646.99.costService)$

Table 4.3 Service advertisement s_{A_2}

$NF \doteq \{< Flights, Travel, "travel \ expertise, \ unrivaled \ excellent \ value..." >\}$
$Op_1 \doteq< findingFlight >$
$I_1 \doteq< (LondonCityAirport, String >, (OrlandoUSA, String), (14/07/07, Date),$
$\quad (17/0707, Date), (1, Adults), (0, Children), (0, Infants),$
$\quad (ReturnTrip, String) >$
$O_1 \doteq< (Ticket, String) >$
$le_{pre1} \doteq (isValid.PaymentCard)$
$le_{eff1} \doteq (\geqslant 0(amount.CreditCard - cost.Ticket))$
$le_{qos} \doteq (\leqslant 12second.qualityResponseTime) \cap (\leqslant 8second.qualityExecuteTime)$
$\quad \cap (\geqslant 8rank.Reliability) \cap (\geqslant 7rank.qualityExceptionHandling)$
$\quad \cap (\geqslant 8rank.qualityAccuracy) \cap (\geqslant 8rank.qualitySecurity)$
$le_{cost} \doteq (= 882.46.costService)$

Table 4.4 Instance of concept Ticket of service ontology so_1

---

instance 25DFWJ **memberOf** Ticket
　　　departure_city **hasValue** "London,UK"
　　　departure_code **hasValue** "EI698"
　　　arrival_city **hasValue** "Orlando,US"
　　　arrival_code **hasValue** "EI699"
　　　departure_date **hasValue** 14-07-07
　　　arrival_date **hasValue** 17-07-07
　　　departure_time **hasValue** 1050
　　　arrival_time **hasValue** 1730
　　　cost **hasValue** 646.99 // including taxes
　　　class **hasValue** "economy"

Table 4.5 Instance of concept TravelVoucher of service ontology so_2

instance 45515G **memberOf** TravelVoucher
bearer **hasValue** "John Smith"
departure **hasValue** "London City Airport"
arrival **hasValue** "Orlando International Airport"
departureDate **hasValue** 140707
arrivalDate **hasValue** 170707
departureTime **hasValue** 1250
arrivalTime **hasValue** 1820
amount **hasValue** 882.46 // including taxes
classLevel **hasValue** 2

Having a user's service requirements and a set of candidate web services, the question of how to discover matching services will be answered in the following section.

4.2.2 Service Matchmaking Algorithms

As already mentioned in Section 4.1, a formalised service requirement is noted as s_R, and a service advertisement as s_A. Matchmaking methods are used to determine the similarity between s_R and s_A in order to find the web service which fulfills the user's requirements best. As we have specified four features of a service in Definition 4.1, the similarity of s_R and s_A depends on these four features' similarity. It can be defined as follow,

$$simService(s_R, s_A) = w_1 \cdot simNF(s_R, s_A) + w_2 \cdot simF(s_R, s_A)$$
$$+ w_3 \cdot simQ(s_R, s_A) + w_4 \cdot simC(s_R, s_A) \tag{4.13}$$

where $\sum_{i=1}^{4} w_i = 1$ and $w_i \in [0,1]$. As the values of $simNF()$, $simF()$, $simQ()$ and $simC()$ are all between 0 and 1, $simService() \in [0,1]$ as well. The weight of each feature should be adjusted according to the specifics of an application domain. Due to the complexity of describing service capabilities, e.g. non-functional information is text-based and functional information includes data types and logical expressions, different parts of service descriptions require different matchmaking algorithm. The $simNF()$, $simF()$, $simQ()$ and $simC()$ will be defined, respectively, in later sections. The $simService(s_R, s_A)$ is also used to rank the final selection set.

4.2.2.1 Model for Service Matchmaking

Based on the service model defined, we propose a flexible service matching mode as illustrated by Fig. 4.3. We assume a service requester to be represented by

$s_R = (NF_R, F_R, Q_R, C_R)$, and a web service by $s_A = (NF_A, F_A, Q_A, C_A)$. For service matchmaking any combination of the set $\{(NF_R, NF_A), (F_R, F_A), (Q_R, Q_A), (C_R, C_A)\}$ can be considered.

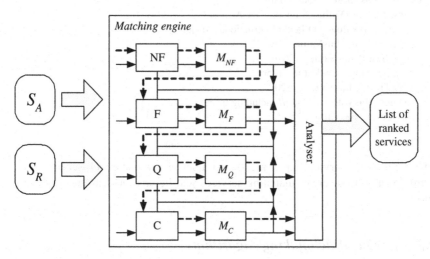

Fig. 4.3 Model of service matching

For instance, if service discovery is performed on the basis of non-functional service information (e.g. service name and text description), only, then the matching process in Fig. 4.3 will follow the lines drawn through from NF to M_{NF} and yield a result. For this case, $simService(s_R, s_A) = simNF(s_R, s_A)$. If service discovery is required to consider first functional and then non-functional service information, then $simService(s_R, s_A) = w_1 \cdot simNF(s_R, s_A) + w_2 \cdot simF(s_R, s_A)$. In case service discovery sequentially performs matching on all four features as the process indicated by the broken line in Fig. 4.3, the result is given by Equation 4.13. The *analyser* is used to deal with the matching results by combining their weights, and its output is a ranked set of matched services.

This model is quite flexible, and sometimes it is very efficient, e.g. it first uses cost to filter candidate web services before going to the details of their functional service information. But obviously, the more information is used by service matchmaking, the more accurate the discovery result is.

Since this service model (by Definition 4.1) has actually three types of data, such as concept or textual description, concept datatype set and logical expressions, different matchmaking methods are needed for different types of data. In the following section, three matching methods are proposed.

4.2.2.2 Concept Matching Algorithm M_C

Taking full account of the concept semantics expressed by the service ontology, we propose a concept matching algorithm M_C of concept $C_1 \in O_1$ and $C_2 \in O_2$. We define

$$simConcept(C_1, C_2) = M_C(C_1, C_2), \tag{4.14}$$

where $M_C() \in [0, 1]$. Let γ be a threshold value for concept similarity. If $M_C(C_1, C_2) \geqslant \gamma$, then we say that these two concepts are similar, also known as $=_s$ equal,

$$C_1 =_s C_2 \tag{4.15}$$

In our service model, the conceptual information can be found in the non-functional properties NF, and by Definition 4.7 it is further elaborated as,

$$simName(sn_A, sn_R) = M_C(sn_A, sn_R) \tag{4.16}$$
$$simCategory(sc_A, sc_R) = M_C(sc_A, sc_R) \tag{4.17}$$
$$simOp(Op_A, Op_R) = M_C(Op_A, Op_R) \tag{4.18}$$

To measure similarity of ontological concepts is actually a very complex task, which is explained in Chapters 5 and 6.

4.2.2.3 Text Matching Algorithm M_T

The only textual information is a short natural-language description of a web service, as $nf{:}sd$ (cp. Definition 4.7). This part of information plays only an auxiliary rôle for the whole discovery process. Nevertheless, when there are enough text-based service descriptions sd_A and service requirements sd_R, matching them is still very helpful.

Table 4.6 Matching algorithm M_T for *serviceDescription*

inputs: Document sd_A, Document sd_R, int N;
output: *simDes*;
Word w; Document d; // the intermediate variables
method1: // df(w), document frequency of word w
float df(w)= the number of times w occurs throughout all documents;
method2: // wf(w,d), word frequency
float wf(w,d) = the inverse to df(w);
method3: // h(w,d), a weight of word w in a document d in N documents
float $h(w, d) = wf(w, d) \cdot \log(N \setminus df(w))$;
method4: // calculating the weighted vectors of sd_A and sd_R
for (word w_i: document d_j)
caculate $h(w_i, d_j)$;
return Vector DesR, DesA;
method5: // calculating the similarity of sd_A and sd_R
double $simDes = \frac{DesR \cdot DesA}{|DesR| \cdot |DesA|}$;
// $simDes \in [0, 1]$

The traditional term frequency-inverse document frequency (TF-IDF, [147]) theory, which originated in the areas of information retrieval and text mining, is re-used here for computing the similarity of service descriptions and denoted as *simDes*. The proposed matching algorithm M_T is defined in Table 4.6 by five methods. The inputs of M_T are the textual service descriptions sd_A and sd_R, and the total number of documents in the calculation space N, the output is the similarity of service descriptions $simDes \in [0,1]$. By *methods1* through *methods4*, weighted word vectors are returned, and then taken by *method5* to compute the final similarity *simDes*.

4.2.2.4 Pairwise (Variable, Type) Matching Algorithm

In the service model defined (cp. Definition 4.1), the input and output declarations of service operations are sets of variable-datatype pairs (vr, dt) (cp. Definitions 4.9 and 4.10). By referring to [162, 183], we propose our definitions for (vr, dt),

Definition 4.1. (Subtype Inference Rules) Let two types t_1 and t_2 be the datatypes of variable vr_1 and vr_2 (noted as $vr : dt$), respectively. Then,

1. type dt_1 is a subtype of type dt_2 (denoted as $dt_1 \preceq_{vt} dt_2$), if this can be deduced by the following subtype inference rules, and
2. two types dt_1, t_2 are equal ($dt_1 =_{vt} dt_2$), if $t_1 \preceq_{vt} dt_2$ and $dt_2 \preceq_{vt} dt_1$ with

    a. $dt_1 =_{vt} dt_2$, if they are identical ($dt_1 = dt_2$),
    b. $dt_1 \mid dt_2 =_{vt} dt_2 \mid t_1$ (commutation), and
    c. $(dt_1 \mid dt_2) \mid dt_2 = dt_1 \mid (dt_2 \mid dt_3)$ (association).

The subtype inference rules read,

r1: $dt_1 \preceq_{vt} dt_2$ if dt_2 is a type variable

r2: $\dfrac{dt_1 =_{vt} dt_2}{dt_1 \preceq_{vt} dt_2}$

r3: $dt_1 \preceq_{vt} dt_1 \mid dt_2$

r4: $dt_2 \preceq_{vt} dt_1 \mid dt_2$

r5: dt_1, dt_1 are sets, $\dfrac{dt_1 \subseteq dt_2}{dt_1 \preceq_{vt} dt_2}$

r6: $\dfrac{dt_1 \preceq_{vt} dt_2, s_1 \preceq_{vt} s_2}{(dt_1, s_1) \preceq_{vt} (dt_2, s_2)}$

r7: $\dfrac{dt_1 \preceq_{vt} dt_2}{SetOf(dt_1) \preceq_{vt} SetOf(dt_2)}$

r8: $\dfrac{dt_1 \preceq_{vt} dt_2}{ListOf(dt_1) \preceq_{vt} ListOf(dt_2)}$

Definition 4.2. (Signature Matching Function) Let fs be a binary variable-type function for $(vr_1 : dt_1)$ and $(vr_2 : dt_2)$ with

$$fs(vr_1 : dt_1, vr_2 : dt_2) = \begin{cases} sub, & dt_1 \preceq_{vt} dt_2 \text{ and } vr_1 =_s vr_2 \\ Sub, & dt_2 \preceq_{vt} dt_1 \text{ and } vr_1 =_s vr_2 \\ eq, & dt_1 =_{vt} dt_2 \text{ and } vr_1 =_s vr_2 \\ disj, & else \end{cases} \qquad (4.19)$$

To discover services, user inputs I_R should be equal or subsume service inputs I_A, and service outputs O_A need to subsume or be equal to the user-expected outputs O_R. By Definitions 4.1 and 4.2 service discovery should satisfy that,

$$fs(I_R, I_A) \in \{sub, eq\} \wedge fs(O_R, O_A) \in \{Sub, eq\} \tag{4.20}$$

where, if every element of I_R is sub or eq an element of I_A, we say that $fs(I_R, I_A) \in \{sub, eq\}$, and analogously for $fs(O_R, O_A) \in \{Sub, eq\}$. Expression (4.20) is also denoted as,

$$(I_R \Rightarrow_{\{sub, equ\}} I_R) \wedge (O_R \Rightarrow_{\{Sub, equ\}} O_A) \tag{4.21}$$

Definition 4.3. (Set Similarity) Let E_a, E_b be variable declarations of service inputs or outputs, and $S(E)$ a set of words in E. The similarity between E_a, E_b is determined by pairwise computation of variable similarity as follows:

$$sim(E_a, E_b) = 1 - \frac{\Sigma_{(u,v) \in S(E_a) \times S(E_a)} simCon(u, v)}{|S(E_a) \times S(E_a)|} \tag{4.22}$$

With the condition of (4.21), the $simInput(I_A, I_R)$ and $simOutput(O_A, O_R)$ are defined in analogy to Definition 4.3, if they are sets of inputs and outputs. For single inputs or outputs, we use the variable similarity as their input or output similarity.

4.2.2.5 Matching Algorithm for Service Pre-conditions and Effects

The P and E of the service model are actually expressed as sets of value constraints on service inputs and outputs.

Definition 4.4. (Semantical Plug-in Service Match) There is a semantical **plug-in match** between service requirements s_R and service advertisement s_A, if

1. their variable signatures match;
2. the set of pre-conditions P_R of s_R logically implies that of P_A of s_A;
3. the set of effects of E_A of s_A logically implies that of E_R of s_R.

Just as [162], we consider for this part of matchmaking the θ-subsumption relation between two logical constraints (cp. Definition 4.5), because it is computationally tractable and semantically sound. Further, it has been proven in the software engineering area [116] that, if the three conditions of semantical matching in Definition 4.4 hold, then the constraints of s_A can directly be used **in place of** s_R, i.e. s_A plugs in s_R.

Definition 4.5. (θ-Subsumption [132]) A clause c_1 is a θ-subsumption clause of c_2 if and only if there exists a substitution θ such that $c_1 \theta \subseteq c_2$. In other words, c_1 is a generalisation of c_2, and c_2 is a specialisation of c_1 under θ-subsumption.

The θ-subsumption inductive inference rule reads:

$$\theta\text{-subsumption: } \frac{c_1}{c_2} \text{ where } c_1 \theta \subseteq c_2$$

For example, $father(X,Y) \longleftarrow parent(X,Y)$, $male(X)$ θ-subsumption $father$ $(jeff, paul) \longleftarrow parent(jeff, paul), parent(jeff, ann), male(jeff), female(ann)$ with $\theta = \{X = jeff, Y = ann\}$.

The clauses of Definition 4.5 are treated here as sets of logical assertions. We combine the notation of Horn clauses to express the set of constraints with a default literal a_0 (if a is used as signature). A general Horn clause is $a_0 \vee (\neg a_1) \vee ... \vee (\neg a_n)$, where each a_i, $i \in \{1, ..., n\}$ is an atom. It is equivalent to $(a_0 \wedge ... \wedge a_n) \Rightarrow a_0$, which can also be written equivalently in the form of an implication, as $a_0 \longleftarrow a_1, ..., a_n$. Examples for the clause a_i are $(\geqslant 8am.DepartTime)$ and $isValid(CreditCard)$ (see Table 4.1). The θ-subsumption between clauses used here is

- $U(a) \leftarrow V(a) \preceq_\theta U(X) \leftarrow V(Y)$
- $U(X) \leftarrow V(X), W(X) \preceq_\theta U(X) \leftarrow V(X)$

Moreover, subsumption between two sets of clauses is defined in terms of subsumption between single clauses. That is, let S and T be sets of clauses, then S θ-subsumes T if every clause in T is θ-subsumed by a clause in S.

Finally, based on Definitions 4.4 and 4.5, we define that a web service **semantically matches** user requirements, if

$$(P_R \Rightarrow_\theta P_A) \wedge (E_A \Rightarrow_\theta E_R) \tag{4.23}$$

Web services, which fulfill the constraints defined by Expression (4.23), will be filtered as candidate results by our discovery method. Therefore, we simplify the definitions as,

$$simPre(P_A, P_R) = \begin{cases} 1, & P_R \Rightarrow_\theta P_A \\ 0, & otherwise \end{cases} \tag{4.24}$$

$$simEff(E_A, E_R) = \begin{cases} 1, & E_A \Rightarrow_\theta E_R \\ 0, & otherwise \end{cases} \tag{4.25}$$

4.2.2.6 Matching Algorithm for Service Quality and Cost M_q

Similar to the P and E, the Q and C of the service model (cp. Definition 4.1) are also sets of constraint clauses. Obviously, service discovery should sufficiently meet that,

$$(Q_R \Rightarrow_\theta Q_A) \wedge (C_R \Rightarrow_\theta C_A) \tag{4.26}$$

In addition, we care more for how these features contribute to service selection by their similarity. Quantitatively measuring the similarity of a set of clauses is quite complex, especially when each clause has different variable types and value scales. This problem will be elaborated by the entire Chapter 7. Here, we only present its definition,

$$simQ(Q_A, Q_R) = M_q(Q_A, Q_R) \in [0, 1] \tag{4.27}$$

Although the cost of a web service, C, is normally defined by a single constraint, it can similarly be calculated by Expression (4.27). Besides, for service selection, we introduce an expression for assessing the approximate degree of service cost. Assuming the same currency unit, the approximate degree of service cost is defined as

$$closeExpCost(C_A, C_R) = \frac{v_A - v_R}{v_R} \tag{4.28}$$

Taking the examples from Tables 4.1, 4.2 and 4.3, we obtain the results shown in Table 4.7.

Table 4.7 Examples of approximate degree of service cost

$closeExpCost(C_R, C_{A1}) = (800 - 646.99)/800 = 0.1913$
$closeExpCost(C_R, C_{A2}) = (800 - 882.46)/800 = -0.1031$

By Equation (4.28), the sign of $closeExpCost()$ indicates the trend of the similarity. Obviously, we expect that a smaller service price is better, that is, the bigger the positive value of $closeExpCost()$ is the better. Thus, in our examples, by price service s_1 is better than service s_2.

Table 4.8 Examples for a goal and a web service

Goal g

$NF = \{< FlightBooking, Travel, "book\ the\ cheapest\ flight" >\}$
$Op_1 \doteq < FlightBooking >$
$I_1 \doteq < (LondonUK, String), (OrlandoUSA, String),$
$\quad (July1407, Date), (July1707, Date), (1, Integer), (0, Integer),$
$\quad (0, Integer), (Return, String) >$
$O_1 \doteq < (Ticket, String) >$
$le_{pre1} \doteq \cap (\leqslant 800cost.Ticket) \cap (\geqslant 8departTime.Ticket)$
$\quad \cap (\leqslant 20departTime.Ticket) \cap (\geqslant 2ndClass.Ticket) \cap (isValid(CreditCard))$
$le_{eff1} \doteq \varnothing$
$le_{cost} \doteq (\leqslant 800.costService)$

Web service s_1

$NF \doteq \{< CheapFlightSearch, Travel, "Searching\ for\ cheap\ overseas\ flights." >\}$
$Op_1 \doteq < cheapFlightSearch >$
$I_1 \doteq < (LondonUK, String), (OrlandoUSA, String), (14 - 07 - 07, Date),$
$\quad (17 - 07 - 07, Date), (1, Integer), (0, Integer), (0, Integer),$
$\quad (ReturnTrip, String), (TravelInsuranceID, String) >$
$O_1 \doteq < (TravelVoucher, String) >$
$le_{pre1} \doteq (isValid(PaymentCard))$
$le_{eff1} \doteq (\geqslant 0(amount.CreditCard - cost.Ticket))$
$le_{cost} \doteq (= 646.99.costService)$

4.2.3   Examples of Matchmaking Algorithms

In the sections above, we have discussed in detail each aspect defined by our service model and their respective matchmaking algorithms. Here, we take our running example from Section 4.2.1.6 to show how these matchmaking algorithms work.

Table 4.8 represents the formal goal g of Table 4.1 and a web service instance s_1 in Table 4.2, which both follow Definition 4.1 of our service model. We provisionally omit the service quality part, which is used for service selection when there are many candidate web services (cp. Chapter 7). Thus, for the similarity of the goal g and web services investigated, we consider Table 4.8.

Concept matchmaking.   Are the concepts used by service name and operation name sufficiently similar? The concept similarity is defined in Equation (6.4.2) and is considered by the service ontology context as M_C (cp. Expression (4.14)). For example, $simConcept(FlightBooking, CheapFlightSearch) = 0.56$. The service category is obviously the same with degree 1.0. According to the result of M_T (cp. Table 4.6), both service descriptions are sufficiently similar with degree 0.746.

Input and output matchmaking.   The variables and data types of inputs and outputs are considered here. For example, I_{s_1} has one more input. Following the subtype inference rules $r7$, $r5$ and $r3$, it holds that $I_g \Rightarrow_{sub} I_{s_1}$ (but not vice versa), and $fs(I_g, I_{s_1}) = sub$, analogously for $fs(o_g, o_{s_1}) = sub$. Moreover, the set of the input variables is the same. With Equation (4.22) we obtain $simI(I_g, I_{s_1}) = 1 - \frac{8+7}{8 \times 9} = 0.76$, and for the output part $simConcept(ticket, travelVoucher) = 0.83$ is calculated in their ontologies in Chapter 5, leading to $simO(O_g, O_{s_1}) = 0.83$.

Service pre-condition, effect and cost matchmaking.   We mainly check their θ-subsumption relation by Definition 4.5. In the pre-condition part, by referring to the ticket instance in Table 4.4, we know that $P_g \Rightarrow_{\theta_p} P_{s_1}$ by $\theta_p = \{departureTime.Ticket = 1050, classTicket = "economy", cost.Service = 646.99, paymentCard = creditCard\}$. Since there is no effect defined in goal g, we say that E_{s_1} θ-subsumes E_g. Table 4.7 lists the approximate cost degrees we have measured.

From the above matchmakings, we deem that web service s_1 meets the user's requirements. If the discovery finds many candidate web services, on the service matching best should be decided by a service selection method (cp. Chapter 6).

4.3   Implementation in WSMX

4.3.1   The Extended SESA

The service model and its matchmaking algorithms were implemented in three new components, and used to extend the semantically enabled service-oriented architecture (SESA), which was detailed in [169, 53]. The extended architecture is called *xSESA* and illustrated by Fig. 4.4.

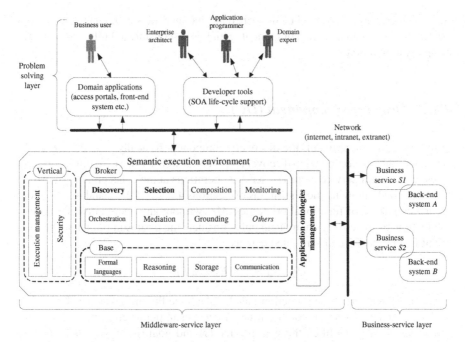

Fig. 4.4 The extended Semantically Enabled Service-oriented Architecture (xSESA)

We choose SESA as implementation context, because it is the most complete implementation of a Service-Oriented Architecture (SOA) with a proposed semantical layer on top of existing service stacks, and at the same time it is compatible to existing industry standards and technologies used within existing enterprise infrastructures. Besides, SESA has been implemented in the Web Service Modeling and Execution Environment (WSMX[6]), which constitutes together with the Internet Reasoning Service (IRS-III[7]) the reference implementation of the Semantic Execution Environment (SEE[8]) as specified by the Organization for the Advancement of Structured Information Standards (OASIS). That is, SESA is quite mature a technology supported by a set of tools, and is easily used to be extended and evaluate our work.

As Fig. 4.4 illustrates, xSESA consists of three main layers, namely business service, problem-solving and middleware services layer. The *middleware services layer* constitutes the Semantic Execution Environment (SEE), shown in the middle of Fig. 4.4. It aims to define any possible task occurring in the context of semantic web services, including discovery, selection, composition, mediation and security. Therefore, at this layer we developed a discovery component (DC) and a selection component (SC) to replace the original service discovery component.

---

[6] http://www.wsmx.org

[7] http://kmi.open.ac.uk/project/irs

[8] http://see.sti-innsbruck.at/

As implementation of the method described in Chapter 5 we supplemented an application ontology management component, which is responsible to build and manage an application ontology.

4.3.2 Discovery Component (DC)

In order to give more detail, the discovery component basically uses the NF or F element of the service model to discover all potential web services. To filter web services with similar service names or service operations at the conceptual level, M_{NF} can be applied involving the algorithms of Sections 4.2.2.2, 4.2.2.3 and 4.2.2.4. For the matchmaking of functional properties, M_F, without any quantitative similarity calculation, the discovery component only considers the θ-subsumption matchmaking of Definition 4.5 and is re-cited here as,

$$(P_R \Rightarrow_\theta P_A) \wedge (E_A \Rightarrow_\theta E_R) \tag{4.29}$$

Therefore, a reasoner-based function matchmaker (see Fig. 4.5) was developed for the DC, which performs pre-condition and effect subsumption between goals/service and services. We simplified the web service (s) and goal (g) as $s = (\Phi_{pre}, \Phi_{eff})$ and $g = (\Psi_{pre}, \Psi_{eff})$, where Ψ and Φ denote the sets of axioms. For example, $\Psi_{pre} \doteq preClause_1 \wedge preClause_2 \wedge ... \wedge preClause_i$, which includes i pre-conditions. The Ψ_{eff}, Φ_{pre} and Φ_{post} are defined similarly.

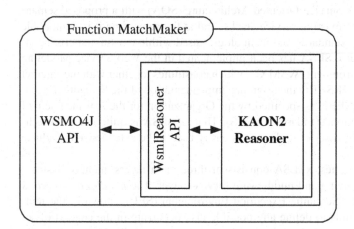

Fig. 4.5 Service matchmaker

Thus, the matching degrees between a goal and a service are defined as follows:

- *Exact(G,S)* iff $(\Psi_{pre} \supseteq \Phi_{pre}) \wedge (\Psi_{eff} = \Phi_{eff})$,
- *Plugin(G,S)* iff $(\Psi_{pre} \supseteq \Phi_{pre}) \wedge (\Psi_{eff} \subseteq \Phi_{eff})$,

- $Subsum(G,S)$ iff $(\Psi_{pre} \supseteq \Phi_{pre}) \wedge (\Psi_{eff} \supseteq \Phi_{eff})$,
- $Intersect(G,S)$ iff $(\Psi_{pre} \cap \Phi_{pre} \neq \emptyset) \vee (\Psi_{eff} \cap \Phi_{eff} \neq \emptyset)$,
- $Fail(G,S)$ iff $(\Psi_{eff} \cap \Phi_{eff} = \emptyset)$.

Obviously, the result provided by the discovery component consists of the web services which have the matching degrees *exact* or *plugin*.

The matchmaker shown in Fig. 4.5 was implemented by a wrapped KAON2 reasoner with WSMO APIs. KAON2 [170] provides support to evaluate logical expressions. The reasoner uses the *WSML2Reasoner* API which extends the *WSMO4J* API providing Java artifacts for WSMO entities. At run time, the matchmaker is invoked as a component of the Web Service Execution Environment (WSMX). As result the matchmaker delivers a list of matched services.

4.3.3 Selection Component (SC)

The selection component is mainly responsible to measure the similarity between a goal instance (a goal with real data) and services at run time. The quantitative similarity of elements NF,F,Q,C of goal and services is measured by this component. According to Definition 4.13, we know that $sim(g,s) \in [0,1]$. In fact, the similarity measurement of the selection component includes matchmaking at the concept, operation and service levels.

The selection component was developed as an Eclipse plug-in in Java in the course of implementing the algorithms for M_C in Section 4.2.2.2, M_T in Section 4.2.2.3 and $M_{\mathscr{Q}}$ in Section 4.27.

4.4 An E-Government Application

In [175] the ontology-dependent and goal-based two-phase approach to discover semantic web services introduced in this chapter was worked out further and utilised in an e-government application. The latter was built using WSMO and following the syntax of WSML to define e-government services and goals. The tree-based algorithm for goal/service discovery was implemented in the discovery component of WSMO. It works on a goal/service ontology, called GSO, defined by ontologising the category concepts of e-government services, and represented as a tree, called GS. The system addresses the problem of how to represent a user's informal needs as formal goals by asking questions while traversing the tree GS thereby deriving a goal template, i.e. a generic description of an objective. Using it, a set of candidate web services is discovered by the mechanism's first phase. Then, in the second phase, a goal instance, i.e. an instantiation of the goal template with concrete requests, is created to finally discover and invoke the web service sought.

Concept mappings between entities occurring in public administration (PA) applications and WSMO elements were investigated with the aim to implement PA services as WSMO services. Defining corresponding ontologies, web services, goals and mediators gave rise to a service model called WSMO-PA. It presents a national as well as a cross-country e-government infrastructure consisting of several member state middlewares integrating all necessary components for service registry, repository and discovery. A community-level execution environment takes care of the communication between them.

In the WSMO-PA service model, GSO plays the rôle of a domain-specific top ontology, which is shared by all other aspects of WSMO-PA. Each object entity of PA services is provided by a corresponding ontology class, through defining its concepts, relations, functions, instances and axioms. In modeling PA services as WSMO web services with capabilities and interfaces, PA pre-conditions were mapped to WSMO pre-conditions, and certain PA pre-conditions, such as the rules of PA services, to WSMO assumptions. Further, the outputs of PA services were mapped to WSMO post-conditions, and their effects straightforward to WSMO effects. With respect to the interfaces, the propagation of the consequences of PA services was covered by orchestration in the WSMO model.

In each member state, a specific web-based user portal serves as front-end application to e-government services invoked by citizens, businesses and service providers to specify user requirements in form of goals as well as to manage user data and the proper services. A discovery component specific to a member state takes user needs to find suitable services by searching repositories of the PA services published and registered. Then, the two-phase discovery procedure works as follows.

1. The user portal's interface guides a user to answer questions, which are generated by traversing the tree GS until a certain node with an associated goal template is reached. Based on the post-conditions defined in that node, a set of services fulfilling the user's needs is discovered.
2. From this set, the user selects one service and enters more data as required by its pre-conditions. A goal instance is instantiated as a real goal with concrete data. Then, the DC uses this instance to discover and invoke the web service sought.

From the users' point of view, the tree GS is used to refine and formalise user needs into formal goals, and from the service providers' point of view, it helps in registering e-government services. With a well-defined tree GS, discovery based on goal templates reduces to a process of finding suitable paths in the tree. The discovery algorithm's inputs include the goal/service ontology, the user's profile as XML document and the user's needs expressed as axioms in first-order logic. The output is a set of goal templates discovered. The matchmaker of the DC compares the pre- and post-conditions of the user's goal with those of the published services.

4.5 Summary

In this chapter we first presented a framework to discover semantic web services, in which service discovery was divided into two phases, one at the static and abstract level and the other at the dynamic and instance level. Based on three assumptions, we then proposed a model for service discovery, which is a tuple $s = (NF, F, Q, C)$ and formally defines all factors necessary to discover services. Going very deep, we further defined for each element of the service model a matchmaking algorithm, and took a running example to show how they work. The details on ontological concept similarity will be further elaborated in Chapters 5 and 6, and the part on service qualities is found in Chapter 7.

Chapter 5
Building Application Ontologies for Semantic Web Services

This chapter presents our second contribution to building application ontologies for semantic web services, which aims to solve the interoperation problems between heterogeneous ontologies in the SWS context. Beginning with the introduction of ontology levels in SWS, we then come to state a building approach for application ontologies.

5.1 Levels of Ontologies and Requirements

Guarino [61] suggested to use ontologies with different levels of specificity in applications. As shown in Fig. 5.1, *Top-level ontologies* describe very general concepts, which are independent of particular problems or domains. *Domain ontologies* describe vocabularies related to generic domains by specialising the concepts introduced in top-level ontologies. *Task ontologies* describe vocabulary related to generic tasks or activities by specialising top-level ontologies. And *Application ontologies* define the concepts in applications often corresponding to rôles played by domain entities while performing certain activities in applications.

In the field of semantic web services, currently we distinguish between *generic ontologies* (instead of *Top-level*), *domain ontologies* and *application ontologies*. Two major *generic ontologies* have been proposed, OWL-S and WSMO, and they use a similar mechanism. These two specifications were compared in [86]. Without lacking generality, our work will follow the WSMO/WSML specifications.

A *Domain Ontology* (denoted as $\mathcal{DO}$) models a specific domain and represents the particular meanings of terms in that domain. An *Application Ontology* (denoted as $\mathcal{AO}$) may involve several domain ontologies in order to describe a specific application. Their relation is defined as $\mathcal{AO} \subseteq \bigcup_{i=1}^{m} DO_i$, $i, m \in N$.

The small piece of ontology attached to a service is defined as a *Service Ontology* (denoted as $\mathcal{SO}$), and it holds $\mathcal{SO} \subseteq \mathcal{AO}$. Any two $\mathcal{SO}$s describing the same domain may intersect or not. Then, an application ontology can finally be defined as $\mathcal{AO} \doteq \bigcup_{i=1}^{n} SO_i$, $n \to \infty$.

X. Wang and W.A. Halang: Discovery and Selection of Semantic Web Services, SCI 453, pp. 63–79.
springerlink.com   © Springer-Verlag Berlin Heidelberg 2013

Fig. 5.1 Different kinds of ontologies and their relationships

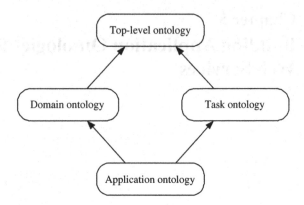

Requirements. Unfortunately, there are no formal standard requirements for building an ontology in the context of semantic web services. Based on our experience, we identify some requirements for the result ontology rendered by a building tool. The result ontology should fulfill:

1. being a globally consistent ontology;
2. having potentially powerful semantic expression and reasoning capability, i.e. the built ontology should be rich in knowledge and useful for further reasoning tasks;
3. being interoperable, robust, usable, adaptable and scalable.

A tool for building ontologies should have some desirable features, namely (1) it can import diverse ontologies, (2) it should be independent of application domains, (3) it should be independent of ontology modeling languages, (4) it should be an automatic tool in order to adapt to the fast increase of web service descriptions, (5) it follow some standards to keep the consistency of ontologies and (6) it should be easy and feasible to use.

5.2 Conceptual Model for Application Ontologies

5.2.1 Definition of Ontologies

In order to formally and concisely define an application ontology, we reconsider the ontology structure proposed by Maedche et al. [98] and expanded by Ehrig et al. [46]. For our purpose, we simplify it (by removing the lexical entries $\mathcal{L}$ and the heterarchy $\mathcal{H}_c$), modify it (by merging relations $\mathcal{F}$, $\mathcal{G}$ into relation $\mathcal{R}$) and extend it with multitype relations, defined as:

Definition 5.1. (Ontology with Datatypes) An ontology with datatypes is a structure $O := (C, \leqslant_C, R, \leqslant_R, \sigma_R, T, A, \sigma_A, I)$ consisting of a set of concepts C aligned in a hierarchical graph with a partial order $\leqslant_C$, a set of binary relations R with $\leqslant_R$, the signature $\sigma_R : R \rightarrow C \times C$, a set of datatypes T with type transition functions, a set of datatype attributes A, the signature $\sigma_A : A \rightarrow C \times T$ and a set of instances I. For a specific relation $r \in R$, we define its domain and its range by $dom(r) := \pi_1(\sigma_R(r))$ and $range(r) := \pi_2(\sigma_R(r))$, where function π_1 gives the domain of relation r as a union of concepts, and function π_2 similarly gives the range of relation r.

No linguistic information is specified at this point.

Definition 5.2. (Ontology Knowledge Base) A knowledge base structure is a quadruple $\mathscr{K}\mathscr{B} := \{\mathscr{O}, \mathscr{I}, inst, instr\}$ consisting of an ontology (with datatypes) $\mathscr{O}$, a set $\mathscr{I}$ whose elements are called instances (which includes the instances defined in $\mathscr{O}$), a function $inst : \mathscr{C} \rightarrow 2^{\mathscr{I}}$ called concept instantiation (for $inst(C)$ one may write $C(I_c)$) and a function $instr : \mathscr{R} \rightarrow 2^{\mathscr{I} \times \mathscr{I}}$ called relation instantiation (to specify domain and range of $instr(R)$ one may write $R(I_1, I_2)$).

This ontology structure will be used by service ontologies, and its relation R has been extended from the single *subclass* relation to multiple relations (such as *similarity*, *hypernym* and *meronym*) with their respective fuzzy weights, which can be used to measure concept similarity.

5.2.2 Semantic Net of Ontological Concepts

The relation R of Definition 5.1 has been extended as a relation set, $R = \{r_h, r_m, r_i\}$, where r_h is the *hypernym/hyponym* (*kindOf* or *subclass*) relation, r_m is the *meronym/holonym* (*partOf*) relation and r_i is the *similar* (including *synonym*) relation. Extended with multiple concept relations, an application ontology cannot be structured as a tree, which is strictly limited by hierarchical concept relations or domain categories. The application ontology should be structured as a semantic net allowing multiple direct parents along each relation in the relation set.

An application ontology $\mathscr{A}\mathscr{O}$ is represented as a graph $\mathscr{G} = (\mathscr{V}, \mathscr{E})$, where $\mathscr{V} = \{c_1, c_2, ..., c_n\}$ is a set of nodes of ontological concepts with their concept type set c_T, and $\mathscr{E} = \{e_1, e_2, ..., e_m\}$ is a set of edges, in which each edge is denoted by a named ($\rightarrow$) binary relation between the two concepts linked. In a data structure for such a semantic concept net, a concept node may maintain separate tables for its attributes, instances and its synonym set (which is a word term synset, e.g. the ones defined in WordNet).

Fig. 5.2 depicts a semantic net with 12 connected nodes and four isolated (grey coloured) ones. The semantic net models three concept relations, with r_h depicted as a solid direct edge, r_m as a dashed direct edge and r_i depicted as a line annotated with a similarity value $sim(C_1, C_2) \in [0,1]$. If $sim(C_1, C_2) = 1$, then the concepts C_1, C_2 are synonyms. In addition, a concept C_i may have an attribute table T_{Ai}, instance table T_{Ii} and synonym table T_{Si}.

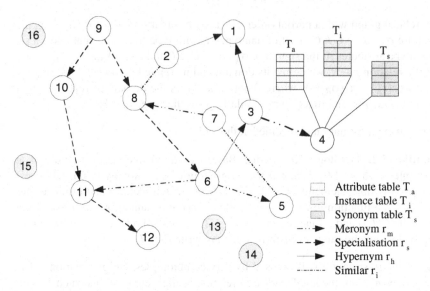

Fig. 5.2 Semantic net of ontological concepts

5.3 Building Application Ontologies

Initially, this work was motivated by considering ontological concept similarity to the end of improving service discovery, which is supposed to find the semantic service matching given service descriptions best. Unfortunately, based on their original limited service descriptions, only, it is nearly impossible to know how similar two concepts are. Through building $\mathcal{AO}$, however, it is possible to predict the similarity of two concepts C_1, C_2, if they are known to relate to a third one C_3, which is added during the building process.

Mining the rich semantics between two ontologies based on term names requires a knowledge base, e.g. a thesaurus, dictionary or corpus. Since a small piece of a service ontology provides only limited information, by itself it is not sufficient for dealing with more complex reasoning and knowledge mining tasks. Moreover, a well structured knowledge base will be very useful to support ontology building and reduce ontology conflicts.

Being quite a mature conceptual thesaurus with hierarchically structured organisation and multiple semantic relations between concepts, WordNet[1] [111] is often used for ontology learning, mapping and merging. As it is widely accepted and recognised, here we employ WordNet as standard for ontology building and alignment as well. When two ontologies have naming or concept relation conflicts, we follow WordNet. We also use WordNet for word sense disambiguation (WSD) in order to extract the synonym and the hypernym/meronym sets of terms. How to carry out WSD is a problem left by [176].

---

[1] http://wordnet.princeton.edu/

5.3.1 Process of Building $\mathscr{AO}$

We assume that there are k semantic services from the same application domain as inputs of the $\mathscr{AO}$ builder. As depicted in Fig. 5.3, the building process is iterative including the following steps.

1. Initialise $\mathscr{AO}_0$ to the empty set, $\emptyset$.
2. Import a service ontology SO_i, $1 \leqslant i \leqslant k$ in an ontology language. Extract the information defined by Definition 5.1 from this SO_i.
3. For each concept of $\mathscr{SO}_i$, extract its relative concept term information from WordNet by the *WCEA* algorithm described in Section 5.3.2, add the retrieved information to $\mathscr{SO}_i$. At this point $\mathscr{SO}_i$ is re-marked as $\mathscr{SO}'_i$.
4. Merge $\mathscr{SO}'_i$ and $\mathscr{AO}_{i-1}$ using the *WOMA* algorithm defined in Section 5.3.3 to obtain $\mathscr{AO}_i$.
5. Process $\mathscr{AO}_i$, including cleaning concept conflicts, updating and storing as described in Section 5.3.4.
6. Increment i and repeat Steps 2 to 5 until $i = k$, then stop.

The output of this process is an application ontology $\mathscr{AO}_k$, which can be used by other service providers or users to create new applications as depicted in Fig. 5.3.

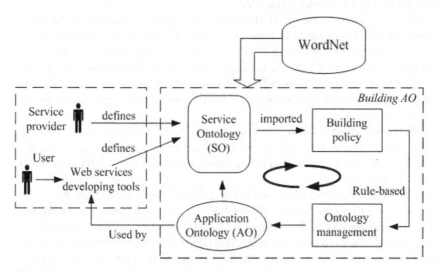

Fig. 5.3 Process of building an application ontology

5.3.2 WordNet-Based Concept Extraction Algorithm

English WordNet[2] is an on-line lexical database of English words. Nouns, verbs, adjectives and adverbs are grouped into sets of cognitive synonyms and near synonyms

---

[2] WordNet versions exist for many languages.

(synsets). Synsets are interlinked by means of conceptual semantic and lexical relations, including *hypernym/hyponym* and *meronym/holonym*. Here we limit ourselves to noun-taxonomy extraction. Although word stems are detected before the merging process, any Part-Of-Speech-related natural language analysis is not carried out.

Assume that an $\mathscr{SO}$ has j concepts $\{C_1, C_2, ..., C_j\}$, $j \in N$. The WordNet-based concept extraction algorithm (*WCEA*) aims to extract additional relevant concept information from WordNet, which has not been defined by $\mathscr{SO}$. It involves the following steps: (1) perform word sense disambiguation on all concepts of $\mathscr{SO}$ (according to Agirre's and Rigau's [3] algorithm), (2) extract two kinds of information of the disambiguated concepts from WordNet, specifically its synset words and its relation words, (3) add these information to the ontology by respectively executing *add...()* operations, e.g. *addSynonym(concept)* and *addRelation(relationR)*.

In the following, we elaborate on two major points of the *WCEA* algorithm.

5.3.2.1 Word Sense Disambiguation

Many words are polysemous or homonymous (i.e. two different concepts are denoted by the same word, e.g. "bank"). In a specific service description, however, a single one of a word's senses is intended.

Word Sense Disambiguation (WSD) [90] is the task of identifying the intended meaning of a given *target word* from its context. In the service ontology $SO_i = \{travel, train\ travel, flight, ...\}$ in Fig. 5.5, for instance, the word *travel* has three noun synsets listed in WordNet 2.1, viz. *(travel, traveling, travelling; Sense #1), (change of location, travel; Sense #2) and (locomotion, travel; Sense #3)*. An issue is which word sense is intended for the terms in this travel service ontology, assuming the meaning is a noun.

Here, we use concept density to consider the WSD algorithm as discussed in [3, 112, 10]. The point is to resolve the lexical ambiguity of nouns by finding the combination of senses from a set of contiguous nouns that maximises the conceptual density among senses [2].

The WSD mechanism is illustrated in Fig. 5.4. Assuming word W has four noun senses and several context words $\{W_1, W_2, W_3, W_4\}$, and each sense of the word belongs to a subhierarchy of WordNet. The dots in the subhierarchies represent the senses of either the word to be disambiguated (W) or the words in the context. Conceptual density will yield the highest density for the subhierarchy containing more senses relative to the total number of senses in the subhierarchy. The sense of W contained in the subhierarchy with highest conceptual density will be chosen as the sense disambiguating W in the given context. Therefore, the sense #2 of Fig. 5.4 would be chosen as the right one.

Given a concept c at the top of a subhierarchy, and given the mean number of hyponyms per node, *nhyp*, Agirre [3] defined the conceptual density for c, when its subhierarchy contains a number m (marks) of senses of the words to disambiguate, to be

Fig. 5.4 Senses of a word in
WordNet [3]

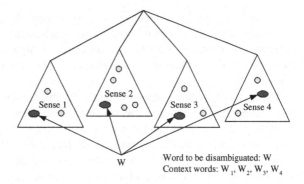

Word to be disambiguated: W
Context words: W_1, W_2, W_3, W_4

$$CD(c,m) = \frac{\sum_{i=0}^{m-1} nhyp^{i^{0.2}}}{descendants_c} \tag{5.1}$$

He found that the best performance was consistently attained when the power of
i was near 0.2. We take the above SO from Fig. 5.4 to explain how our WSD
works. First, according Agirre's algorithm, we adjust it to $SO = \{train\,travel, travel, flight\}$, with $travel(3)$ as the *target word* in the middle (the number in parentheses
indicates how many senses the word has), the other two words $\{train(6), flight(9)\}$
as window context symmetrically ordered on both sides (meanwhile we remove the
duplicate word "travel" from "train travel") and 3 as the size of the target word
window.

Table 5.1 Partial lattice for the example concept

Flight_#9
⇒ trip
⇒ journey, journeying
⇒ **travel_#1**, traveling, travelling
⇒ motion, movement, move
⇒ change
⇒ action
⇒ act, human action, human activity
⇒ event
⇒ ...

caravan, train, wagon train_#3
⇒ procession
⇒ group action
⇒ act, human action, human activity

Next, the WSD algorithm starts. Initially, it represents all words of the word window into their lattices, with their senses and hypernyms (Step 1). Two concepts of
our example with their senses are illustrated in Table 5.1. Then, the conceptual density of the word window's concepts is computed referring to Equation (5.1) (Step 2).

In Table 5.1, the $(act, human\ action, human\ activity)$ is the common sense level found, and the subhierarchy of word "*flight*" leads to the highest conceptual density: 0.25806. It is, therefore, selected as a concept with high density (Step 3) and, in turn, the senses in this subhierarchy are selected as the correct senses of the respective words (Step 4).

The WSD algorithm goes on to deal with the remaining word windows, if any, and continues to disambiguate words (by repeating the loop from Step 2 to 4 until no more disambiguation can be processed). The result is presented in Step 5. In our case, two words are finally disambiguated: $\{travel\#1, flight\#9\}$.

Sometimes this algorithm fails to disambiguate all concepts of a word context. This problem can often be resolved by repeated invocation of the WSD algorithm during the ontology building process.

5.3.2.2 Retrieval Operations on WordNet

After concept disambiguation, the meaning of some concepts is specified. At this point, operations *addSynonym( )* and *addRelation( )* can be considered. For instance, "*travel#1*" has a synset of "traveling, travelling", and these words will be added to the synonym table of the concept "travel".

Concept association is the criterion for selecting the kind of interconcept relation to be added. If two concepts are related by one of these interconcept relations in the service ontology, but no such relation is defined in the service ontology yet, then add it to this service ontology. In our example, there is already a *subclass* relation between $flight\#9$ and $travel\#1$ in the service ontology, so $addRelation(r_s)$ is not necessary here.

5.3.3 WordNet-Based Ontology Merging Algorithm

5.3.3.1 Preprocessing

When importing an $\mathscr{SO}_j$ and merging it into the current $\mathscr{AO}_i$, the existing structural description of the global ontology merging process [100] can be re-used and extended by several preprocessing steps as follows:

1. Import a $\mathscr{SO}_j$ and clean words (word stemming and removing redundant words), transform $\mathscr{SO}_j$ into a semantic net, organise concept attributes or instances into respective tables.
2. After WSD, extract concept information of $\mathscr{SO}_j$ from WordNet by executing the operations *add(synonym)*, *add(attribute)*, *add(instance)*, *delete(relationR)* and *add(relationR)* as necessary. For example, concept $C_k \in SO_j$ has three synonym words found in WordNet. Then, these three words will be added into C_k's synonym table T_s by *add(synonym)*. Instances are added to the attribute table T_a and the instance table T_i in a similar fashion.

3. Align and process $\mathscr{SO}_j$. If C_k has two subconcepts, C_{k1} and C_{k2} in SO_j, but in WordNet C_{k2} is subconcept of C_{k1}, then the respective modify operations are $add(r_s)$ between C_{k1} and C_{k2}, and $delete(r_s)$ between C_k and C_{k2}. The conflict relations with WordNet are removed at this stage.

5.3.3.2   Merging Process

After preprocessing, we obtain a new ontology $\mathscr{SO}'_j$. According to the structure of its semantic net, the concepts of $\mathscr{SO}'_j$ are orderly stored as $SO'_j = \{C_{x_1,0}, C_{x_2,0}, ..., C_{x_m,0}, C_{y,z}, ..., C_{0,r_1}, ..., C_{0,r_n}\}$ (where x, y, z, r are variables and y, z are non-zero). Concepts are ordered according to the increasing *in-degree* (as left subscript) and, in case of equality, ascending *out-degree* (as right subscript). In this order are, first, the concept nodes with 0 out-degree (including isolated ones), behind them stand their parents and the rest is deduced by analogy until the concept nodes with 0 in-degree.

Then, the WordNet-based ontology merging process is defined as follows:

1. Copy current $\mathscr{AO}_i$ to $\mathscr{AO}_{i+1}$.
2. Merge concepts of $\mathscr{SO}'_j$ into $\mathscr{AO}_{i+1}$ (denoted as $\mathscr{AO}_{i+1} = \mathscr{AO}_i \cup SO'_j$).
3. Clean conflicts in $\mathscr{AO}_{i+1}$, such as name conflicts or taxonomy conflicts.

5.3.3.3   Rule-Based Merging Mechanism

We assume that C_1 and C_2 are two concepts of ontologies $\mathscr{AO}_i$ and $\mathscr{SO}'_j$, and $C_1 \in \mathscr{AO}_i, C_2 \in \mathscr{SO}'_j$. Several rules must be followed during the merging process. We present examples of the rules taken from the ontologies *travel1* and *travel2*[3]. The rules are :

Rule 1:   If C_1 and C_2 are lexically identical, and if they have concept types and are of the same type, then they are identical. Execute Rule 6.

For instance, concept *"Travel"* with the data type of _String_ is defined in two ontologies.

Rule 2:   If C_1 and C_2 are lexically identical, and if they have different types and there is no transition function between their types, then they are different concepts. Copy and add concept C_2 to ontology $\mathscr{AO}_i$ after re-naming. Or else check Rule 4 if they are possibly identical to other terms.

For instance, two concepts have the same name *"Area"* with different concept types *"Square Meter"* and *"Acre"*. Then, the merged concept with its type would be $(Area, \{SquareMeter, Acre\})$, a function $(1 \, acre = 4,047 \, squaremeters)$ had been defined previously. Such axioms are normally given on concepts in WSMO ontologies.

Rule 3:   If C_1 and C_2 are lexically identical, and if the attribute tables $T_{A2} \cap T_{A1} = \emptyset$, then they are not identical. C_2 is added as a new concept after re-naming, see Section 5.3.4.

---

[3] http://deri.org/iswc2005tutorial/ontologies/

Rule 4:   If C_1 and C_2 are not lexically identical, and if $T_{A2} \neq \emptyset$ and $T_{A2} \subseteq T_{A1}$, then C_1 and C_2 have a *meronym* relation (r_m): Create a new relation between C_1, C_2 and add it to $\mathscr{AO}_i$. If $T_{A2} = T_{A1}$, then C_1, C_2 are synonyms and Rule 7 is executed. If $T_{A2} \cap T_{A1} = \emptyset$, then C_2 is added as a new concept.

For instance, concept C_1, *"Customer"*, has an attribute set $\{firstname\_String,$ $lastname\_String, street : \_String, city : \_String, zipcode : \_String, country : \_String\}$, and concept C_2, *"Name"*, has attributes $\{first : \_String, last : \_String\}$. By our *possibleName* policy[4], a *meronym* relationship is discovered between them.

Rule 5   is similar to Rule 4 but applies to the concept instance table instead of the attribute table.

Rule 6:   If C_1 and C_2 are identical, then merge the two concepts.

For instance, *"Date"* is an identical concept in two ontologies, then keep the one of $\mathscr{AO}_i$.

Rule 7:   If C_1 and C_2 are synonyms, then merge concept C_2 into C_1's synonym table in $\mathscr{AO}_i$.

Concepts *"Ticket"* and *"TravelVoucher"* in Fig. 5.5, for example, are considered synonyms by our algorithm. In the first round of the mapping, identical attributes are found, and in the following rounds each time just one attribute (in big, bold font, whose implied type is another concept) is expanded to look for further matching possibilities. The mapping relationship could be $m : n$, e.g., *"toFrom"* is mapped to the set $\{departure\_city, departure\_code, arrival\_city, arrival\_code\}$. By such a kind of expanded checking policy, the potential similarity of the attributes shown in small, bold font can also be discovered. Here, concept *"TravelVoucher"* is finally found to be similar to *"Ticket"* and added to its synonym table.

Rule 8:   If a transitive relation relates C_1 to C_2, and C_2 to C_3 in $\mathscr{SO}'_j$, but C_1 is *related* to C_3 by the same relation in $\mathscr{AO}_i$, then delete the relation between C_1 and C_3 in $\mathscr{AO}_i$ and specify the new relations: $createRelation(r, C_1, C_2)$, $createRelation(r, C_2, C_3)$ ($r \in R$ being the relation set defined in Section 5.2.2).

This rule is used to deal with concept relations. In Fig. 5.6, for instance, there is a *subclass* ($r_s \in R$) relation between $hotel \subset Travel$ (in ontology $O1$) and $hotel \subset accommodation \subset Travel$ (in ontology $O2$). In the end, we merge them by adding $hotel \subset accommodation \subset Travel$ to $O1$.

5.3.3.4   Some Merging Operations

Based on the above rules, specific operations should be executed. The following list contains some of the defined operations, where C denotes *Concept* and R denotes *Relation*.

---

[4] This policy is an optimisation strategy. It first collects all term names of two ontologies. During concept merging, it then checks any possible name case in order to find the best match, e.g. concept *"Name"* will update its attribute set to $\{firstname : \_String, lastname : \_String\}$, because *"firstname"* and *"lastname"* are the possible names.

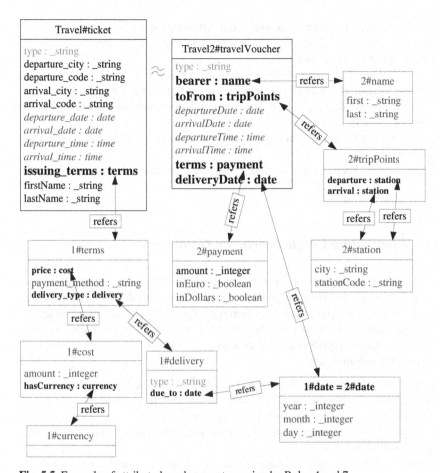

Fig. 5.5 Example of attribute-based concept merging by Rules 4 and 7

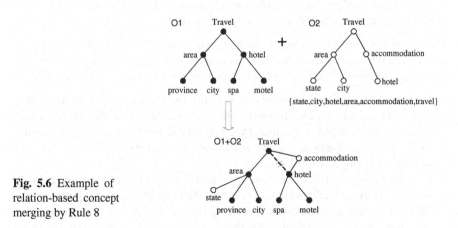

Fig. 5.6 Example of relation-based concept merging by Rule 8

- *copy(C)*: copies a concept's attributes and instances from its ontology.
- *addConcept(C,AO)*: after *copy()*, assigns a newly created concept to $\mathscr{AO}$'s namespace and adds it into $\mathscr{AO}$.
- *copyRelation(r,C)*: copies all the *R* relations of *C* from its ontology, e.g. a *sub-Concept* relation (r_s) between *C* and *C'* is copied via *copyRelation(r_s,C)*.
- *createRelation(r,C_1,C_2)*: establishes a new relation (subclass, partOf, similar) between two concepts.
- *addRelation(r,C_1,C_2,AO)*: after *copyRelation()* or *createRelation()*, adds this relation into *AO*.
- *merge(C_1,C_2)*: keeps C_1 and copies C_2, merging C_2's attributes and instances into C_1.

5.3.4   Eliminating Conflicts between Ontologies

After merging and copying concepts, it is necessary to check the consistency of the generated ontology and to clean up the conflicts that occurred. Some related work [6, 62, 140] specifies types of ontology conflicts. Ram [140] characterises two levels of semantic conflicts (at *data* and *schema* level) between ontologies. Conflicts may arise in the context of SWSs as listed in Table 5.2. The conflicts checked here and their general resolutions are:

- *Concept entities conflicts*: Are there more identical concepts? Are there essentially different concepts using the same name? Are there different data types defining one concept, and so on? The general principle is to merge the identical concepts into one, re-name the concept, keep the diversity of types and create type transition axioms, if any.
- *Concept relations conflicts*: Check if there is a relation cycle in a new ontology (by the loop check algorithm of [40]), check if there is a relation conflict *with its original ontology* (by the unification algorithm of [40]) and, if necessary, add newly calculated relations, e.g. synonym, similarity.

These conflicts mostly arise due to a lack of merging rules or standards. Fortunately, they can be resolved by using WordNet as concept structure standard and the merging Rules 1–8 above. The following examples explain the conflicts of Table 5.2.

- Concept data type/unit conflicts: e.g. in [140], the *Pima* county database stores yearly tax amounts for each property, while *Pinal* county stores monthly tax rates as a percentage of property value. Type/unit transition axioms must be manually defined to avoid conflicts, e.g. the transition function between *amount* and *percentage*, referring to Rule 1.
- Concept Name conflicts: e.g. concept *"bank"* in two tourism ontologies, one means a financial bank and the other means the bank of a river. Such name conflict are automatically resolved by taking one of its hypernyms as prefix and re-naming it, i.e. *"riverBank"*. Another kind of example is that {*zip, zipCode, zipCode1, code, code1*} or {*CalcDisTwoZips Km, findZipCodeDistance*} will be

Table 5.2 Ontology conflicts on *data* and *schema* levels

Conflict	Description
Data value conflict	Different interpretations of the meaning of data instance values
Data representation conflict	Similar objects are described by different data types or data format representations
Data unit conflict	Use of different measurement units
Naming conflict	Labels of schema elements (i.e. entity classes, relationships and attributes) are somewhat arbitrarily assigned by different application designers (homonyms and synoyms)
Schema isomorphism conflict	Same concept (entity class) is described by a dissimilar set of attributes (i.e. the same concept is represented by a number of different attributes) or is not set operation-compatible
Generalisation conflicts	Different design choices for modeling related entity classes
Aggregation conflicts	When an aggregation is used in one ontology to identify a set of entities in another one
Schematic discrepancies	When the logical structure of a set of attributes and their values belonging to an entity class in one ontology are organised to form a different structure in another one

seen as synonyms in a certain *Routing* service. This conflict can be resolved by using a concept synonym table, which includes possible synonyms extracted from WordNet or synonyms defined a priori by a certain concept similarity algorithm defined in Chapter 6.

- Taxonomy conflicts: e.g. in [45], a service selling flight tickets can use an airport code ontology which defines *city* as a property of the *AirportCode* concept, and at the same time use a different ontology for countries, in which *city* is a concept and not a property. These two ontologies are not used in a consistent way. For such a case, WordNet's concept structure is treated as standard, and *city* is taken as a concept.
- The remaining conflicts in Table 5.2 are about ontology schema and structure, which can be handled by applying merging Rules 1–8.

5.4 Experimental Prototypes

In the course of the project "KnowledgeWeb" supported by the European Union, we developed a set of *WSAO* building tools[5] based on two designs. One is to build a uniform application ontology through importing and merging published service ontologies, presented in this chapter. The other is to improve semantic service selection and discovery by taking such application ontologies as a knowledge base ($\mathcal{KB}$ by Definition 5.2). The second design is explained in Chapter 6.

The experimental prototype designed, named *WSAO* Studio, consists of a set of Eclipse plug-ins, which are the tools *WSD*, *OntoMerge* and *OntoBuilder*. As shown

---

[5] http://wsao.deri.ie

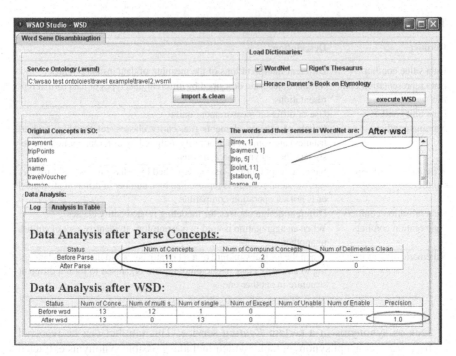

Fig. 5.7 WSAO Studio — Word Sense Disambiguation

in Fig. 5.7, the *WSD* plug-in imports a service ontology[6], e.g. the *"travel2.wsml"* ontology, and parses them with *wsmo4j*[7] to yield 11 concepts. By using WordNet as dictionary and executing our *WSD* algorithm, the concepts with their calculated senses are listed as results. At the analysis panel we see, that when parsing concepts, two compound concepts are found, *"tripPoints"* and *"travelVoucher"*. Then, the number of concepts changes to 13, in which 12 concepts are polysemous. Finally, all words found their senses, and the precision is 1.0. Similarly, we import ontology *"travel1"*, the precision obtained is 0.82, i.e. 10 of 12 concepts found their senses.

Fig. 5.8 is a screen snapshot of the ontology merging tool. It imports two ontologies and executes *WSD*, respectively, then merges these two ontologies into one. Table 5.3 presents the statistics: initially there are 22 concepts, 80 attributes, 4 instances and 2 *subConcept* relations in total. The merged ontology compacts this to 17 concepts, 54 attributes, 4 instances and 5 relations — including the one $r_s(time, date)$ derived from WordNet, and two attributes, *name* and *station*, added to the concept *customer*. The similarity of each pair of concepts can be calculated (cp. Chapter 6), e.g. the similarity of *"ticket"* and *"travelVoucher"* is 0.88359.

The $\mathcal{AO}$ builder is shown in Fig. 5.9. It can build a totally new ontology by importing two single service ontologies or a batch of services under a directory. The

---

[6] Currently, it only imports files written in the *WSML* language.
[7] http://wsmo4j.sourceforge.net/

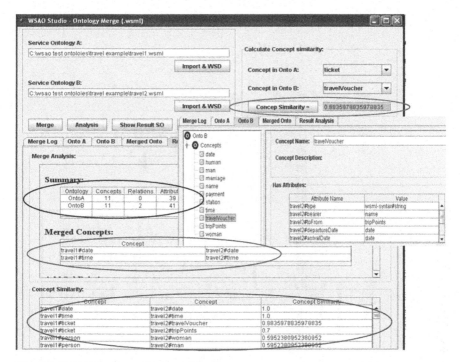

Fig. 5.8 WSAO Studio — Ontology Merging

Table 5.3 Data analysis of ontology merging

Ontology	Number Concepts	Number Attributes	Number Instances	Number Relations	Merged Concepts	Copied Concepts
travel1	11	39	2	0	–	–
travel2	11	41	2	2	–	–
merged	17	54	4	5	4	7

result ontology is visualised in a graph panel (so far it uses the WSMO visualisation component). It can build a new ontology based on an existing ontology, or import a batch of service ontologies from a file directory. As illustrated in Fig. 5.9, the $\mathscr{AO}$ is built by merging four different travel-related ontologies.

Since this ontology building algorithm checks concepts, concept attributes, instances and relations (of three types), the time complexity is $O(N^2)$ with respect to the number of ontology terms. For numbers of terms less than 50, the time requirements show an almost linear increase. The time to merge the *travel1* and *travel2* ontologies is 0.547 sec, and the time to merge *airReservation* and *carRental* is 0.662 sec on an IBM laptop computer running at 2.16 GHz and having 1 GB RAM. More evaluation details of these tools are given in Chapter 8.

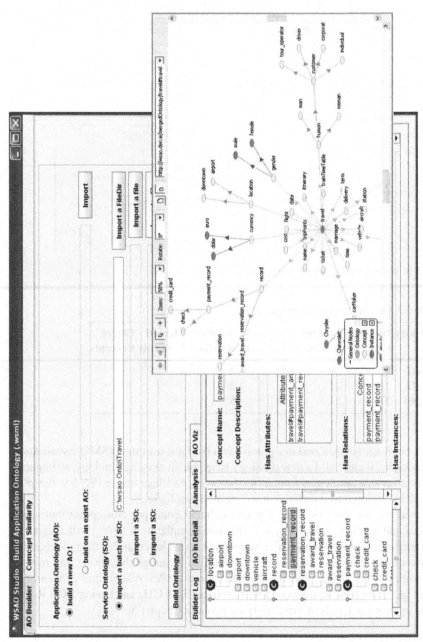

Fig. 5.9 WSAO Studio — Ontology Builder with the resulting travel ontology

5.5 Summary

In this chapter a novel approach for building application ontologies in the SWS context was proposed, using a rule-based algorithm for automatic merging of semantic service ontologies. The contributions of this approach can be summarised as: (1) it considers word sense disambiguation of ontology concepts before merging ontologies, (2) WordNet is not only used to extract more concept relations (e.g. synonyms, hypernyms or meronyms) to discover implicit information among ontological concepts, it is also taken as merging standard, (3) a rule-based ontology merging algorithm is proposed, which considers both the merging of concept relations and of concept content, (4) a complete prototype was built to evaluate the algorithms and (5) a set of realistic tools, named as *WSAO* Studio, was developed to implement these algorithms. In the next chapter, it will be explained how the measurement of service concept similarity and of service similarity can be based on application ontologies.

Chapter 6
Ontology-Based Service Selection

After the introduction of building application ontologies for a kind of web services in the previous chapter, we go on in this chapter from the semantic web services' point of view to investigate how to calculate concept similarity in such application ontologies. Moreover, the problem of measuring concept similarity between two heterogeneous service ontologies is transferred here into the one of computing semantic concept distance in one ontology.

In detail, after graphically representing the ontological concepts and concept relations in a fuzzy-weighted semantic net, then an original ontological concept similarity algorithm is proposed in this chapter, which takes multiple concept relations into consideration. Finally, it is shown how such concept similarity promotes the computation of similarity between semantic web services.

6.1 Motivation

From Chapter 4 we know, that similarity of semantic web services is mostly measured by the conceptual overlap of service capabilities, including non-functional information, e.g. service name, service categories and service qualities, and functional properties, i.e. service operations. That is, concept similarity is the core issue.

However, it is quite complex to measure the similarity of service concepts. In order to illustrate this, Table 6.1 gives an example of five web services related to postal codes (these services were obtained from the *Semantic Web Services Repository* of the University College Dublin[1]. These web services are used to look up zip codes or calculating distances between two places given their zip codes.

Taking *sws4* and *sws5* of Table 6.1 as examples, we assume that *zip* and *code* have a certain similarity in this specific application. By comparing service names and service operation names, these services can be regarded as similar from the signature level. In addition, by matching their operations and inputs/outputs, we can conclude

---

[1] http://moguntia.ucd.ie/repository/

X. Wang and W.A. Halang: Discovery and Selection of Semantic Web Services, SCI 453, pp. 81–95.
springerlink.com   © Springer-Verlag Berlin Heidelberg 2013

Table 6.1 Examples of semantic web service descriptions

---

ws1: **ServiceName:** Zip_Code_Lookup
    operation: GetZipByCityState
        input: StateCode:string, CityName:string
      output: Zip:string

ws2: **ServiceName:** CityStateByZip
    operation: GetCityStateByZip
        input: ZipCode:string
      output: GetCityStateByZipResult:ArrayOfString

ws3: **ServiceName:** Distance_between_two_zip_codes
    operation: CalcDistTwoZipsKm
        input: Zip_Code_1:string, Zip_Code_2:string
      output: CalcDistTwoZipsKmResult:double

ws4: **ServiceName:** zipCodeService
    operation1: findZipCodeDistance
        input: code1:int, code2:int
      output: findZipCodeDistanceResult:double
    operation2: findZipDetails
        input: code1:int
      output: findZipDetailsResult:ArrayOfString

ws5: **ServiceName:** Zip4
    operation1: FindZipPlus4
        input: Address:string, City:string, State:string
      output: Zip:string
    operation2: FindCityState
        input: Zip:string
      output: City:string, State:string

---

that services *sws4* and *sws5* are similar — both services can provide information on a city according to a given zip code. Furthermore, if we assume that a machine can understand that there is some similarity between $\{zip, ZipCode, code, Zip\_Code\_1, code1\}$ or $\{CalcDisTwoZipsKm, findZipCodeDistance\}$, then we could derive that *sws1:operation* is similar to *sws5:operation1*, *sws2:operation* is similar to *sws5:operation2* and *sws4:operation2* as well as *sws3:operation* is similar to *sws4: operation1*. As it becomes obvious from this example, ontological concept similarity based on both syntactic and semantic approaches is the basis for service similarity.

From the related work mentioned in Chapter 3, we know that the measurement of concept similarity is still a problem for semantic web services with the above concept features. Moreover, a service does not only use single domain ontology to define its applications, but may in fact use several domain ontologies. For example, a travel service may involve concepts from the finance and the transportation domain. The relations among concepts are turning very complex. Therefore, an application ontology for a kind of semantic services cannot be simply structured as a traditional tree-like network, which emphasises only the *subtype* or *is-a* relations between concepts of a domain, but must have the form of a hybrid semantic net including multiple kinds of concept relations. That is the main reason why the existing concept similarity algorithms are of limited usefulness in semantic web service scenarios.

Fortunately, application ontologies according to Chapter 5 solve this problem. When such an application ontology is being built, three factors are fully considered, namely (1) determining the correct meaning of the concepts used by web services, (2) retrieving more synonyms and concept relations (e.g. hyponyms and meronyms) from WordNet based on concept senses and (3) merging service ontologies based on concept types, attributes and relations. Therefore, such an application ontology has richer semantic information than before, which is a good knowledge basis to measure concept similarity.

6.2 Ontology-Based Selection Framework for Semantic Web Services

Corresponding to Section 4.1, on one side a user states his/her service requirements as s_R (say, a *WSMO-goal* in the WSMO framework), and at the other side there are many available candidate services, $s_{A_1}, s_{A_2}, ..., s_{A_n}$, published (say, *WSMO-web services*). As shown on the right-hand side of Fig. 6.1, a service selection engine compares each pair (s_R, s_{A_i}), $i \in \mathcal{N}$ by matching/filtering their *non-functional* (including service qualities) and *functional* service requirements with service advertisements, and ranks them by semantic service similarity.

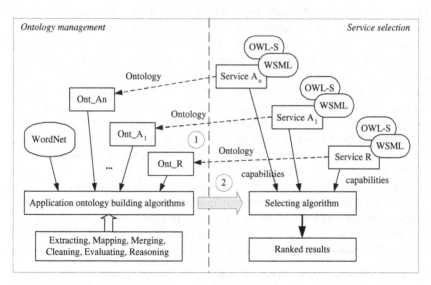

Fig. 6.1 Ontology-based semantic service selection framework

As we have already stated, the current service selection approaches are mostly based on the assumption that all service descriptions use a single ontology. This assumption is unrealistic in real-world scenarios where services are free to use their own ontologies. Therefore, in Fig. 6.1, the left-hand side presents a component for building an application ontology as our solution to the integration of services with different ontologies (cp. Chapter 5). Moreover, the output of Step 2 is an application ontology ($\mathcal{AO}$), which is then inversely used by the *selection algorithm* as a common ontology. In fact, the *Application ontology building* component is transparent to service providers and users, and service descriptions would deem using one single ontology. Our approach of building application ontologies reduces misunderstandings between service descriptions and prevents difficulties in service interoperation. Service concept similarity is calculated within such application ontologies.

6.3 Ontological Concepts in the Context of Semantic Web Services

Before we consider concept similarity, let us first introduce some prior knowledge on the ontology and concept features of SWSs.

6.3.1 Concept Features

From the examples of web service descriptions in Section 6.1, we summarise three kinds of special concept features appearing in SWSs as follows:

- There are two types of concepts. Some are formal and the other are composite ones combined by several concepts, which are not formal vocabularies indexed by a dictionary, e.g. WordNet. For such composite concepts, e.g. *findZipCodeDistance* and *CalcDistTwoZipsKm*, the traditional methods, e.g. the *edit distance* of strings, are not applicable to measure their semantic similarity. Therefore, measuring similarity of composite concepts is a major problem.
- A concept has a single sense in a specific service application description, e.g. *code* in our above service examples does not mean *cipher* nor *computer code*, but is actually used as synonym of *zip code*. Therefore, word sense disambiguation is necessary before considering concepts similarity, which was addressed in Chapter 5;
- Information shared by concepts should be more than what is represented in small pieces of service ontologies as currently used by semantic service descriptions. Further, concept relations should be more than just the *isA* relation. Therefore, mining implied information and concept relations is necessary and contributes to assess service similarity.

Therefore, in this chapter two types of concept similarity for formal concepts and for composite concepts, respectively, will be addressed.

6.3.2 Concept Definition in WSMO

An ontology formally defines concepts and concept relationships in a machine-understandable way with the goal to enable sharing and re-use of knowledge. According to the definition given in Section 5.2.1, ontological concepts are described by their datatypes, attributes, instances and the relations with other concepts, all of which are to be considered when measuring concept similarity.

Table 6.2 Ontology definition in WSMO

class Ontology
hasConcept **ofType** (0 *) concept
hasRelation **ofType** (0 *) relation
hasAxiom **ofType** (0 *) axiom
hasInstance **ofType** (0 *) instance
class concept **subConceptOf** wsmoElement
hasType **ofType** (1) datatype
hasAttribute **ofType** (0 *) attribute
hasDefinition **ofType** logicExpression
multiplicity = single-valued

Without loss of generality, we specify our defined ontology as a WSMO ontology as indicated in Table 6.2, which is implemented with the WSML-DL language. From Table 6.2 we see that an ontology generally uses a set of *axioms* to capture its actual statements, which are defined by a formula in a logical language; and the *hasDefinition* is also a logical expression which can be used to formally define the semantics of the concept. In Table 6.3, for instance, concept *Child* is defined with an axiom, where *dc* is the identifer of the imported ontology[2].

Table 6.3 Example for an ontological concept in WSMO

concept Child **subConceptOf** Human
nfp   dc#relation **hasValue** ChildDef   **endnfp**
axiom ChildDef
nfp
dc#description **hasValue** "Human being not older than 14"
endnfp
definedBy
forall ?x (?x **memberOf** Human and ageOfHuman(?x,?age)
and ?age $\leqslant$ 14 **implies** ?x **memberOf** Child).

6.4 Semantic Concept Similarity

In the following two sections, we propose our concept similarity algorithms in the context of semantic web services.

---

[2] http://purl.org/dc/elements/1.1#

6.4.1   Semantic Net with Multiple Concept Relations

As mentioned in Section 5.1, an $\mathscr{AO}$ may involve several $\mathscr{DO}$s to describe a certain application. Therefore, $\mathscr{AO}$ cannot intuitively be structured as a hierarchical tree-like structure, which can only express the strict superconcept and subconcept relations of single domains, but as a semantic net with any possible concept relations.

Here, we extend the *relation* of Definition 6.2 with the four concept relations *specialisation/generalisation* (or *"is-a"*), *meronym/holonym* (or *"is-a-part-of"*), *hyponym/hypernym*(or *"is-a-kind-of"*) and *similar* (which contains *synonymous*). Although *generalisation* generalises *holonym* and *hypernym*, we still keep the *generalisation* relation, because it is inherited from its original service ontology during ontology merging processes, and to specify them is out of our scope. The relations of *holonym*, *hypernym* and *synonymous* are the new concept relations retrieved from WordNet during the building process of $\mathscr{AO}$. Here, the proposed relation *similar* ranges in $(0,1]$. It contains the relation *synonymous*, and is used when some similar knowledge is presupposed or received from previous work to optimise the current similarity calculation.

Using multiple types of concept relations in such a semantic net, the $\mathscr{AO}$ is represented as a fuzzy-weighted graph [35] $\tilde{G} = (V, E, \tilde{c})$, which consists of a set V of nodes v_i and a binary relation E of edges $e_k = (v_i, v_j) \in V \times V$, where $head(e_k) = v_i$ and $tail(e_k) = v_j$. Each edge associates with a weight, say $\tilde{c}_{i,j} = \tilde{c}(v_i, v_j)$, $\tilde{c}_{i,j} \in \tilde{c}$. Then, an edge with its relation $r_k \rightarrow v_i \times v_j$, $r_k \in R$, is represented as a tuple $< v_i, v_j, r_k, \tilde{c}_{ij} >$. For our case, there are four types of concept relations specified as $R = \{r_s, r_i, r_m, r_h\}$, and their respective relation weights are $\tilde{c} = \{\tilde{c}_s, \tilde{c}_i, \tilde{c}_m, \tilde{c}_h\}$. For example, in Fig. 6.2 the triple $< 8, 6, s, \tilde{c}_s >$ on edge $e_{8,6}$ shows that concepts C_8 and C_6 have a *specialisation* relation with weight $\tilde{c}_s$.

Also, a path p from node c_i to c_j is defined in this semantic net as $p = \{c_0, c_1, c_2, ..., c_l\}$ (where c_0 denotes c_i as the start of the path, and c_l denotes c_j) with l edges. Therefore, the summed weight of this path is defined as $wp = \sum \tilde{c}_k l_k$, where $k \in \{s, h, m, i\}$, $\tilde{c}_k$ is the weight value of relation k, and l_k is the number of relations k of this path.

All relation weights are between 0 and 1, which are the parameters given by users or derived from experiments with concrete data sets from the application domain. Then, we regard the summed weight of a path between two concepts as measure of the similarity of two concepts.

In Fig. 6.2 a semantic net with 16 connected and four isolated (grey) nodes is depicted. The semantic net has four concept relations, *specialisation* (depicted as a solid direct edge, r_s), *meronym* (depicted as a dashed direct edge, r_m), *hyponym* (depicted as a dotted direct edge, r_h) and *similar* (r_i) depicted as a line annotated with a similarity value $sim(c_1, c_2) \in (0,1]$. If $sim(c_1, c_2) = 1$, then two concepts c_1, c_2 are synonyms; if $sim(c_1, c_2) = 0$, then they are completely dissimilar. Any concept c_i may have attribute table T_{Ai}, instance table T_{Ii} and a synonym table T_{Si}.

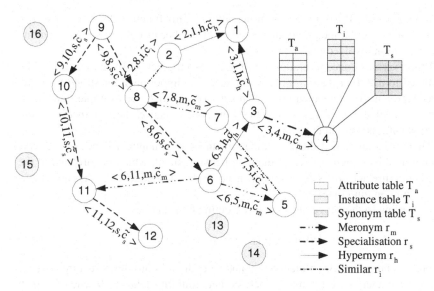

Fig. 6.2 Conceptual ontological semantic net

6.4.2 Similarity Algorithm for Formal Concepts (OCSA)

In the context of research work on semantic web services such as [87, 92, 144, 25], concept similarity is calculated both from the point of view of ontology structure (i.e. the *conceptual distance*, which is the shortest path between two linked concepts, as $Distance_s$), and from the point of view of ontology information content (i.e. the more properties two concepts have in common, the more closely related they are, as $Distance_c$).

Agreeing with Rada's statement [135], *"distance is a metric on the sets of concepts (single and compound) defined on a semantic net"*, we propose a novel algorithm to calculate concept similarity as follow:

$$M_C(c_1, c_2) = Distance(c_1, c_2) \qquad (6.1)$$
$$Distance(c_1, c_2) = w_s \cdot Distance_s(c_1, c_2) + w_c \cdot Distance_c(c_1, c_2) \qquad (6.2)$$

where c_1 and c_2 are two concepts of the semantic net, and w_s and w_c are, respectively, the weights of the structure similarity and the content similarity of these two concepts with $w_s + w_c = 1$.

From the structure point of view, concept similarity does not any longer depend on the path length (l), the concept depth (h) and the concept density (d) of its ontology as proposed by [91], since the application ontology is a graph and not a hierarchical tree. Specifically, we define the concept similarity to relate to path length and concept depth, so that it is denoted as $sim(c_1, c_2) = g(l, d)$. As a metric, the function $g(x, y)$ must have properties such as (1) $g(x, x) = 0$ (reflexivity), (2) $g(x, y) = g(y, x)$

(symmetry) and (3) $g(x,y) \geqslant 0$ (non-negativity). This function does not, however, satisfy the triangular inequality discussed by [135], because there is more than one single relation involved in our case.

When various concept relations exist in a semantic net, different tags should be assigned to different links in order to indicate the importance (or strength) of the links between the connected concept nodes. In Fig. 6.2, for example, the tag $< 8, 6, s, \tilde{c}_s >$ of the edge $e_{8,6}$ shows that concepts c_8 and c_6 have a *specialisation* relation with the weight $\tilde{c}_s$.

Then, we assume that the semantic relatedness or semantic distance of concepts is measured by weight values of the path between them, which is unrelated to the number of directed edges connecting concept nodes. The edge length has no semantic meaning in our context.

According to the above discussion, we define $Distance_s$ as

$$Distance_s(c_1, c_2) = g(g_1(l), g_2(d)) \tag{6.3}$$

where g_1 and g_2 are two functions related to path and ontology structure, respectively. Characteristic for semantic nets is that similarity tends to 0 when the path between two concepts tends to infinity, and otherwise similarity tends to 1. Therefore, we take an exponential function to define,

$$g_1(l) = e^{-wp} = e^{-\Sigma \tilde{c}_k l_k}. \tag{6.4}$$

If only one relation is considered in Equation (6.4), then $g_1(l) = e^{-\tilde{c} \cdot l}$, which is consistent with [46].

Furthermore, we propose a simplified method to compute the shortest path. If the similarity of two concepts on a path has been evaluated to be smaller that a certain value, such a path is not a valid one for our approach. Therefore, we define that a concept similarity value should not be less than a threshold τ. If we set $\tau = 0.2$, then by Equation (6.4) $e^{-0.8 \cdot l} = 0.2$, and $l \doteq 2.012$ when $\tilde{c}_s = 0.8$ (the similarity weight is assumed to be 0.8). Here, we assign $l = 3$ corresponding to human intuition, i.e. if two concepts are similar, the distance of them is not too long.

If there is no path between nodes c_i and c_j in such a semantic net, but there is a common superconcept that subsumes these two concepts, then the conceptual similarity can be measured based on the maximum depth (h) to their common parent node. This also possible in our approach, when we consider four different relations. Thus, referring to [46], we define,

$$g_2(h) = \frac{e^{\beta h} - e^{-\beta h}}{e^{\beta h} + e^{-\beta h}} \tag{6.5}$$

To summarise, based on Equations (6.3), (6.4) and (6.5), we have,

$$Distance_s(c_1, c_2) = \begin{cases} e^{-\Sigma \tilde{c}_k l_k} \cdot \frac{e^{\beta h} - e^{-\beta h}}{e^{\beta h} + e^{-\beta h}}, & c_1 \neq c_2, \\ 1, & otherwise \end{cases} \tag{6.6}$$

where $\beta \geqslant 0$ is the parameter used to scale the contribution of depth h in a concept hierarchy. Equation (6.6) considers the shortest path and the concept depth to their common superconcept, and is consistent with the previous related work.

From the point of view of information content held by concepts, if concepts have rich information, such as concept attributes, instances and given synonyms, then the concept distance can be measured as follows,

$$Distance_c(c_1, c_2) = \frac{|c_1 \cap c_2|}{|c_1 \cap c_2| + \gamma |c_1 \setminus c_2| + (1 - \gamma)|c_2 \setminus c_1|} \tag{6.7}$$

where $|c_1 \cap c_2|$ is the intersection of two concepts indicating the concepts' common characteristics and $|c_2 \setminus c_1|$ is their difference, $|\ |$ is the cardinality of a concept information set and γ ($0 \leqslant \gamma \leqslant 1$) is a weight that defines the relative importance of the concepts' non-common characteristics.

Summary. So far, we have presented the similarity measurement of formal concepts by considering concept relations and content (including attributes and instances), see Equations (6.2), (6.6) and (6.7). Based on this, in the following section we address the similarity measurement of composite concepts.

6.4.3 Similarity Algorithm for Composite Concepts

Composite concepts very often appear in the description of semantic web services. A composite concept is actually a concept consisting of several other incomplete formal concepts, e.g. $\{CalcDisTwoZipsKm, findZipCodeDistance\}$ are two composite concepts extracted from two pieces of zip-related web service descriptions.

Here, the similarity of two composite concepts can also be measured by their concept distance, denoted as $sim(c_i, c_j) = dis(c_i, c_j)$. Similar to [135], we distinguish between three basic cases to calculate concept distance:

- If concepts C_i, C_j are single concept terms, e.g. Zip and $Code$, then

$$dis(C_i, C_j) = \text{the shortest distance of } C_i \text{ and } C_j. \tag{6.8}$$

- If concepts C_i and C_j are composite or conjunctive concepts, e.g. $CalcDis$ $TwoZipsKm$ and $findZipCodeDistance$, then $dis(C_i, C_j)$ is defined as,

$$dis(x_1 \wedge ... \wedge x_k, y_1 \wedge ... \wedge y_m) = \frac{1}{km} \sum_{i=1}^{k} \sum_{j=1}^{m} dis(x_i, y_j), \tag{6.9}$$

where $\{x_1, x_2, ..., x_k\}$ are the subconcepts of concept C_i, and $\{y_1, y_2, ..., y_k\}$ are the subconcepts of C_j, e.g. $CalcDisTwoZipsKm = \{Calculate, Distance, Two, Zip, Kilometer\}$.

- If only one concept C_i or C_j is a composite concept, then $dis(C_i, C_j)$ is defined as,

$$dis(x_1 \wedge ... \wedge x_k, C_j) = \frac{1}{k} \sum_{m=1}^{k} dis(x_m, C_j). \tag{6.10}$$

where the concept $C_i = \{x_1, ..., , x_k\}$ is assumed to be a conjunctive one.

6.5 Semantic Service Similarity

Based on the foundation of ontological concept similarity and reconsidering the service Definition 4.13 in Section 4.2.2, we specify an ontology-based similarity of semantic web services as follows,

$$sim_{Service} = \sum sim_{Concept} + \sum sim_{Operation} \tag{6.11}$$

where $sim_{Concept}$ is the sum over the similarities of all ontological concept-related service descriptions, especially their non-functional properties, and $sim_{Operation}$ is the sum over the similarities of services' conceptual operation parameters. Referring to Definitions 4.16 and 4.22, the service similarity of Equation (6.11) is extended as follows,

$$\begin{aligned} sim_{Service} = {} & e_1 \times simName + e_2 \times simCategory + e_3 \times simDes \\ & + e_4 \times simOp + e_5 \times simInput + e_6 \times simOutput \end{aligned} \tag{6.12}$$

where $\sum_{i=1}^{6} e_i = 1$. The approach taken is fully based on the built application ontologies, and its detailed evaluation is presented in Chapter 8.

6.6 Algorithm and Experimental Prototype

The above concept similarity algorithms are presented by Table 6.4, which includes several methods aiming to calculate the similarity of any two concepts of $\mathscr{AO}$.

If a concept is represented in the semantic net of an application ontology as an isolated node, then its concept similarity is 0.0 by Method 1. If two concepts are not isolated, then Method 2 is used to count any possible paths between them, and Method 5 is applied to calculate the distance of each path. Methods 3 and 4 are used when two concepts have a common parent and, finally, return the distance between them. From the point of view of ontology structure Method 6 is to compute the summed distance, and the shortest path is taken as their distance. Method 7 calculates the concept distance from the point of view of concept information, and Method 8 yields the final similarity of two concepts.

In the subalgorithm, a method for computing all paths between two concepts is listed, say $path(c_1, c_2)$. This method goes through the loop from line 10 to line 14 to track and mine any possible indirected relations between concepts c_1 and c_2 and its

Table 6.4 Concept Similarity Algorithm

Algorithm: Semantic Net-based Similarity *OCSA*

Inputs: concepts c_1, c_2, and application ontology $\mathscr{AO}$;
Output: concept *similarity s* of concepts c_1 and c_2;

Initialisation: relation weights $\tilde{c}_s, \tilde{c}_h, \tilde{c}_m \in (0,1)$, $\tilde{c}_i = 1.0$;

Methods:
1. *isolated*(c);
   // If one concept is an isolated node, then s=0.0.
2. *path*(c_1, c_2);
   // Calculate paths between c_1 and c_1.
3. *hasParent*(c_1, c_2);
   // If there is no path between c_1 and c_2, but they have
   // a common parent, the depth h is the maximum
   // number of edges to the parent concept.
4. *similarityD*(c_1,c_2,h);
   // Calculate concept similarity by depth h.
5. *similarityR*(c_1,c_2,$path_i$); ($i = \{1,..,k\}$)
   // If there are k paths between concepts c_1 and c_2,
   // then calculate the similarity of each path.
6. *similarityS*(c_1,c_2)=max(*similarityR*)$\times$*similarityD*;
7. *similarityC*(c_1,c_1);
   // Check the similarity of c_1 and c_2 by their
   // common attributes and instances.
8. $s = w_s \cdot$ *similarityS* $+ w_c \cdot$ *similarityC*;

Subalgorithm of ***path***(c_1,c_2);
9. if (*similar*(c_1,c_2)==true) then path[i];
10. rs1 [ ][ ]=*relationSet*(c_1);
11. rs2 [ ][ ]=*relationSet*(c_2);
12. if (*linkBySimilar*(rs1,rs2)=true) then pathS[j];
13. else if (*linkByRelation*(rs1,rs2)==true)
14.   then pathR[k];
15. pathNumber=i+j+k;

---

relay concepts (which are the ones collected into the array list $rs1[\,][\,]$ and $rs2[\,][\,]$). Basing on an optimised policy, $l = 3$, the loop then only needs to run once.

This algorithm was implemented in our experimental prototype (cp. Section 5.4). Fig. 6.3 shows part of an experimental scene of building a zip-related web service application ontology. The input is a zip-related service ontology, and by referring to WordNet there are 2 more meronym relations and 11 more synonym relations finally retrieved. Fig. 6.4 shows the ranked results of concept similarity calculation simulated by the algorithm. It imports the relations of concepts (where concepts are indexed by keys), and currently considers the shorted path of concepts involving different relations. All relation weights are tested between $0.8 \sim 0.95$ by referring to [73].

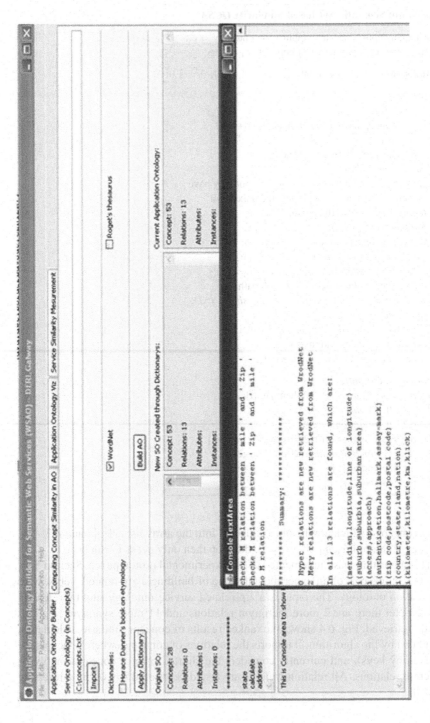

Fig. 6.3 Snapshot of building a zip-related $\mathscr{AO}$ using WordNet

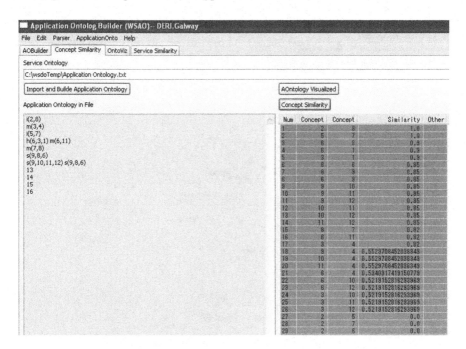

Fig. 6.4 Ranked result of concept similarity

A visualised example of an application ontology is shown in Fig. 6.5. It contains four types of concept relations. Attached to concepts, the attributes are depicted as rectangles and the instances as triangles.

6.6.1 Discussion of Experimental Parameters

From Section 6.4.2 we can see that all six types of parameters contribute to the measurement of concept similarity. These parameters are the weights for balancing structure information and content information, w_s and w_c, the relation weights $\tilde{c}$, the path length l, the depth parameter β and the information content parameter γ. Although the values of these parameters needed to be given through training a large number of real application scenarios, here we can only discuss some general restrictions between these parameters.

1. Assuming that all relations have the same weight $\tilde{c}$ ($\tilde{c} \in (0,1)$) in Equation (6.4), then the restriction of the parameter l and the relation weight is simplified to $e^{-l\cdot\tilde{c}} = s$, where $l \in \{1,2,3,...\}$ and $\tilde{c}, s \in [0,1]$. Generally, the lower l is, the smaller is s, and the bigger $\tilde{c}$ is, the bigger is s.
2. Assuming that all relations have the same weight ($\tilde{c}$) and $l = 3$, the correlation of parameters $\tilde{c}$ and s is $s = e^{-3\cdot\tilde{c}}$. Thus, when $s \geqslant 0.8$, then $e^{-\tilde{c}\cdot3} \geqslant 0.8$, and the value of $\tilde{c}$ should be $\tilde{c} \leqslant 0.693$.

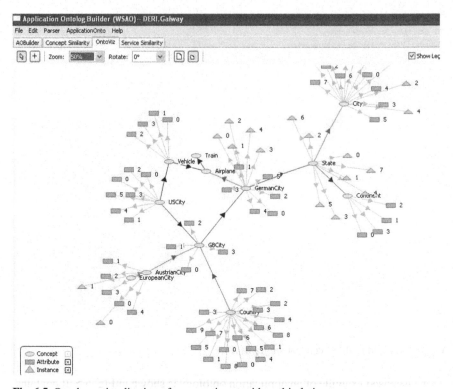

Fig. 6.5 Ontology visualisation of a semantic net with multirelations

3. Similarity, assuming $h = 3$ in Equation (6.5), the correlation of parameters β and
   s is $s = \frac{e^{3 \cdot \beta} - e^{-3 \cdot \beta}}{e^{3 \cdot \beta} + e^{-3 \cdot \beta}}$. Thus, when $s \geqslant 0.8$, the value of β is $\beta \geqslant 0.366$.
4. The time complexity of the algorithm in Table 6.4 tightly relates to path length
   l and in-degree (d) of a concept node in the semantic net, thus it is $O(d^l)$. If
   there are n concepts in the semantic net, then the maximum in-degree of a node
   is $(n-1)$, so that the time complexity is $O((n-1)^l)$.

The above discussion of the application ontology model can be summarised as follows: (1) the model tightly relates to the features of thesauri or dictionaries used,
e.g. WordNet, the richer or better structured a thesaurus is, the higher is the quality
of the application ontology built, and (2) more powerful reasoning capability can
render the application ontology algorithm more effective.

6.7 Summary

Motivated by scenarios to measure the similarity of semantic services, this chapter focused on the measurement of ontological concept similarity of application

ontologies built. The main contributions of our concept similarity approach are: (1) it extends concept relations from single to multiple types enabling to represent more semantics between concepts, (2) it measures semantic concept similarity by employing concept knowledge from ontology structure and concept information content and (3) it solves the problem of calculating composite concept similarity. In conclusion, the concept similarity problem as mentioned in Chapter 4 was addressed in detail.

Chapter 7
QoS-Based Selection of Semantic Services

Quality of Service (QoS) is a very important issue in the field of semantic web services, and the selection of semantic services highly depends on matching service qualities. Specifying the properties of service qualities is very complex, however, not to mention a combined evaluation of all QoS properties. In this chapter we strive to make up for this shortcoming. In continuation of Section 4.2.2.6, this chapter elaborates on the topic of matching service qualities and further proposes a QoS-based approach to service selection.

7.1 Motivation

Most tasks of SWSs, such as automatic discovery, selection, composition, invocation or monitoring of services are tightly related to the Quality of Service of services. As part of the service description, QoS is an especially important factor for service discovery and selection [93]. When a set of similar services is yielded by a discovery process after a series of matching non-functional and functional properties of services, then it depends mainly on the service qualities to decide which one will finally be invoked.

In the literature, this issue has not been thoroughly addressed yet, due to the complexity of QoS metrics. Sometimes, the quality of a service is dynamic or unpredictable. Moreover, most of the current work focuses on the definition of QoS ontologies, vocabularies or measurements, and lacks a uniform evaluation of qualities. The defined service model has briefly specified how QoS contributes to semantic service discovery by Definition 4.26 in Chapter 4. But to measure how close (or similar) two sets are with respect to QoS will be elaborated in this chapter.

Without loss the generality, also here we take WSMO to define our QoS ontology specific to quality metrics, value attributes and their respective measurements. Furthermore, we propose an algorithm to normalise different quality attributes, providing for a dynamic and fair evaluation of web services.

X. Wang and W.A. Halang: Discovery and Selection of Semantic Web Services, SCI 453, pp. 97–107.
springerlink.com

7.2   Similar Work and Discussion

Although we have outlined the related work on service QoS in Section 3.3, some discussion of similar work needs to be added here in order to make this issue more clear because of the relative independence of this subject.

Some similar work focuses on the development of QoS ontology languages and vocabularies, as well as on the identification of various QoS metrics, and their measurements with respect to semantic services. For example, the authors of [125, 164, 63, 160] created QoS ontology models and proposed QoS ontology frameworks aiming to formally describe arbitrary QoS parameters. Their on-going work shows that they did not consider QoS-based service matching, yet. Additionally, the authors of [93, 115, 109, 149] attempted to conduct proper evaluations and proposed QoS-based service selection, but failed to present fair and effective evaluation algorithms.

The work of [93] is quite similar to ours as well. There are, however, some differences with our approach: (1) the measurement of linguistics-based qualities is not considered by them, (2) their algorithm uses average ranking, neglecting nuances in different quality properties, (3) a possible maximum value is used by them to normalise the QoS matrix, although such kind of value is worth deliberating, and (4) upon analysing experimental data, after normalisation their final result set is,

$$
\begin{array}{cccc}
 & q_1 & q_2 & q_3 & q_4 \\
Q_1 & \begin{pmatrix} 0.769 & 1.429 & 1.334 & 1.111 \\ 0.946 & 0.571 & 0.666 & 0.889 \end{pmatrix}
\end{array} \tag{7.1}
$$

That is, they normalised two sets of service qualities Q_1 and Q_2 with four different quality attributes (q_1, q_2, q_3, q_4) by quality types (according to the columns of the matrix above). Obviously, for the same quality property their normalisation can measure similarity, but they cannot make a fair evaluation of all qualities, because they neglect the metrics to have different ranges. One quality attribute has even a higher weight, while its real impact is decreased by its smaller value. Therefore, here we propose to normalise each quality metric to values between 0 and 1 by specifically defined measurements, which are fair to each quality metric.

7.3   QoS Ontology and Vocabulary in WSMO

In WSMO, quality properties are classified as parts of the non-functional information in web service descriptions. For example, *accuracy, availability, financial, network-related QoS, performance, reliability, robustness, scalability, transactional* and *trust* are defined with short notes. But such definitions are insufficient to express QoS attributes in a flexible way.

Thus, we introduce a new concept to the WSMO ontology for defining service qualities, the QoS concept. Furthermore, we define a QoS model following the same syntax to extend the WSMO model. This QoS model can be referred to by the *web service* and *goal* entities, and the quality factors can adequately be considered during the process of service selection. We also specify a QoS upper ontology named WSMO-QoS. It is a complementary ontology providing detailed quality properties about web services. Developers benefit from the WSMO-QoS for QoS-based matchmaking and measurement.

7.3.1 QoS Ontology and Vocabulary

Based on [164, 115, 125] we define a new concept class *QoS* (cp. Table 7.1), which is a subclass of the concept *nonFunctionalProperties* already defined by WSMO. Class *QoS* can be cited by the classes *webService* and *Goal*. Note that the current conceptual model of WSMO remains unchanged, as we just refine the class *nonFunctionalProperties*.

Table 7.1 QoS ontology in WSMO

class QoS **subClassOf** nonFunctionalProperties
hasMetricName **ofType** string
hasValueType **ofType** valueType
hasMetricValue **ofType** value
hasMeasurementUnit **ofType** unit
hasValueDefinition **ofType** logicalExpression
multiplicity = single-valued
isDynamic **ofType** Boolean
isOptional **ofType** Boolean
hasTendency **ofType** {small, large, given}
isGroup **ofType** Boolean
hasWeight **ofType** string

From Table 7.1 we know that each QoS metric is generally described by *MetricName, ValueType, Value* (given or calculated at service run time), *MeasurementUnits* (e.g. €, milliseconds), *ValueDefinition* (how to calculate the value of this metric) and *Dynamic/Static*. For the purpose of QoS-based selection, there are four additional features defined, namely *isOptional, hasTendency, isGroup* and *hasWeight*. A simple interpretation of each qualities' properties is given below.

- Types of the attribute *valueType* may be *linguistic, numeric* (int, float, long), *Boolean* (0/1, True/False) or others. Therefore, there will be different forms of preprocessing according to different value types.
- The attribute *MetricValue* defines a metric's value which is either a real number or a string (i.e. *'calculate'*). If *MetricValue='calculate'*, then this attribute should refer to its *valueDefinition* for a dynamic value calculation.

- The attribute *MeasurementUnit* specifies the concrete unit of each quality metric with possible types, such as $Unit = \{€, milliseconds, percentage, kpbs, times,$ $...\}$. In addition, class *Unit* has a conversion function between different measurement units, e.g. to transform *seconds* to *milliseconds*.
- Attribute *hasValueDefinition* is either a WSMO logical expression (cp. [24]) or the string *'NULL'*. If *hasValueDefinition='NULL'*, then this value definition cannot explicitly be extracted from the context of service descriptions, and must dynamically be derived from the service provider. In this case, this quality attribute must be dynamic, i.e. *isDynamic=True*.
- Attribute *isDynamic* is used to describe the nature of a quality, as *static* or *dynamic*. The values of static qualities are given a priori, and can directly be used during the selection process. If *isDynamic = True*, the quality metric must dynamically be invoked and obtained from its service provider, and its values must be calculated at run time.
- Attribute *isOption* is used to annotate a quality property, say q_k. If *isOption* $= 0$, then q_k is a necessary property and $q_k \in \mathcal{Q}_N$, where $\mathcal{Q}_N$ is the set of necessary qualities. In detail, this property is further described in Section 7.4.
- The object attribute *hasTendency* represents the tendency of a value expected from the user's perspective. For example, the price of a web service is expected to be as low as possible, so that it has *hasTendency='low/small'*. On the contrary, the quality of a service's *security* should be as high as possible, i.e. *hasTendency='high/large'*. When *hasTendency='given'*, the user expects the value of this quality to be as close to the given value as possible. Also, in an inquiry of service quality, *hasTendency=*{*low/small, high/large, given*} is denoted respectively as $\{\geqslant, \leqslant, =\}$ when they are used for their *MetricValue*.
- Attribute *isGroup* indicates if this quality attribute is defined by a group of other qualities or not. For example, *security* is composed of *nonRepudiatior, DataEncryption, Authorisation, Authentication, Auditability* and *Confidentiality* [164]. Hence, *isGroup = True* means that the group value must be calculated first at the preprocessing stage.
- Finally, *hasWeight* is an attribute denoting the weightiness of the property, especially when synthetically measuring several metrics. Here, we define the weight value either to range in $[0, 10]$ or to be *'NULL'*. Different end users have different weight values for their service requirements. Apparently, this attribute is useroriented, thus it is only used by the WSMO *goal*, which describes users' desires. In the description of WSMO web services, its value is set as *'NULL'*.

When a QoS profile is parsed in the service selection process in order to obtain a metric's value for which *hasMetricValue='calculate'* holds, the attribute of *hasValueDefinition* is checked first to determine how to calculate it. If *hasValueDefinition='NULL'* and *isDynamic=1*, then the invocation function is to inquire the current value, otherwise an error is encountered. If *isDynamic=0*, then its corresponding *hasMetricValue* is an existing value, or else again an error occurs.

In [88, 103, 141] all possible QoS properties of web services were defined, including performance, reliability, scalability, capacity, robustness, exception handling, accuracy, integrity, accessibility, availability, interoperability, security,

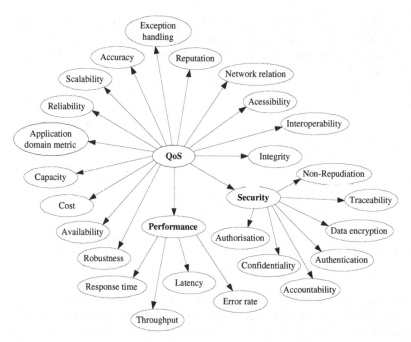

Fig. 7.1 QoS ontology and vocabulary

network-related requirements and so on. In Fig. 7.1, a simple view of QoS vocabulary is given, which consists of many general QoS properties and the scalable domain-specific QoS subset used, e.g., to define the hotel category for a kind of hotel web service. Concrete measurement of qualities was defined and discussed by the references mentioned above and is out of this chapter's scope.

7.4 QoS-Based Selection of Semantic Services

For better understanding, we revisit the content of Chapter 4. The user provides his/her requirements (including non-functional, functional and quality properties) for an expected service, which are formalised as $s_R = (NF_R, F_R, Q_R, C_R)$ (cp. Section 4.2). On the other hand, there may be thousands of available services published in either a service repository or a kind of peer-to-peer service environment. The advertisement of a service s is similarly defined as $s_A = (NF_A, F_A, Q_A, C_A)$.

The first filter of web services is to discover any web services which meet the requirements functionally. This stage is based on service NF and F information, and we assume that m web services are found, viz. $resultSet_1 = \{s_1, s_2, ..., s_m\}$, $m \in \mathcal{N}$.

The second filter synthetically considers all quality properties to select from *resultSet*$_1$ the web service matching best. This matchmaking takes place between the pair of the QoS requirements Q_R and the quality profile Q_A of a candidate web service $s_A \in$ *resultSet*$_1$, as illustrated in Fig. 7.2.

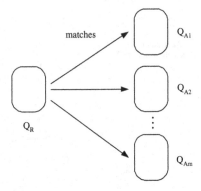

Fig. 7.2 QoS-based selection of services

For the purpose of matchmaking, a QoS selection model is defined, in which metrics are defined both from the perspectives of users and of service providers. Here, we elaborate the content of Section 4.2.1.3. We assume that $\mathscr{Q} = \{q_1, q_2, ..., q_i\}$, $i \in \mathcal{N}$ and $\mathscr{Q}_I$ is the service quality set, and specify that,

- $\mathscr{Q}_N$ is a set of service qualities, which are all necessary to each web service. It is initialised by the system, and denoted as $Q_N \subseteq \mathscr{Q}_I$;
- $\mathscr{Q}_O$ is a set of service qualities, which contains the optional quality properties of web services, as $\mathscr{Q}_O = \mathscr{Q}_I \setminus \mathscr{Q}_N$; and
- Q_D is the default set of service qualities. It takes effect when the user does not explicitly give any quality requirements, i.e. when $\mathscr{Q}_R = \emptyset$ and $\mathscr{Q}_D \subseteq \mathscr{Q}_I$, then $\mathscr{Q}_D$ is taken as $\mathscr{Q}_R$ (denoted as $\mathscr{Q}_R = \mathscr{Q}_D$), where $\mathscr{Q}_R$ are the user's quality requirements. Generally, we assume $\mathscr{Q}_N \subseteq \mathscr{Q}_D$.

There are two reasons to distinguish between different QoS sets. One is to free customers from multifarious definitions of their quality requirements, which sometimes need professional knowledge. For example, customers may not know the meaning of *availability* of web service, but they may really need to take care of such a requirement. For these cases, customers can only provide qualities based on their personal opinions, and the indispensable part may very likely be filled by the default quality set. The other reason is that only few service qualities defined by $\mathscr{Q}_N$ are useful in efficiently performing QoS-based service selection.

7.4.1 QoS-Based Selection Modes

Based on the above analysis, three selection modes are defined with respect to $\mathscr{Q}_R$ as follows:

1. Default mode: if $\mathcal{Q}_R \neq \emptyset$, then $\mathcal{Q}_R$ is re-defined as union of the original user requirements and the default service qualities, $\mathcal{Q}_R := \mathcal{Q}_R \cup \mathcal{Q}_N$;
2. Totally based on the user's requirements, and $\mathcal{Q}_R \neq \emptyset$;
3. Totally based on default service quality set, when $\mathcal{Q}_R = \emptyset$.

In order to efficiently and flexibly perform service selection from the users' perspective, only several qualities are defined in the necessary set, viz. $\mathcal{Q}_N$={*cost, responseTime, reliability, accuracy, security, reputation*}, and similarly $\mathcal{Q}_D$={*cost, responseTime, reliability, accuracy, security, reputation, executionTime, execption-Handling*}. Of course, this definition is extensible and changeable for any specific application.

7.4.2 Collection of QoS Metrics

Although the metrics of service quality properties were specified by Definition 4.11, the approaches of collecting their values need to be clarified as follows,

- directly obtained from the service descriptions, e.g. sometimes the price of invoking a service is given a priori;
- performing a simple calculation based on its definition expressed in the service description;
- collecting values through active monitoring, e.g. the property *execution duration* as explained in [93];
- dynamically inquiring the current server;
- periodically updating the quality values for a statistical purpose using the system log; and
- obtaining customer feedback on the quality of service invocation and execution, e.g. the *reputation* of web services, cp. [93].

In fact, the collection of quality properties at run time is dynamic, unpredictable and even quite difficult. The values of quality metrics can approximately be characterised as,

- numerical metric, denoted by numbers with different value ranges;
- ordinal and linguistic metric, denoted by terms from an ordered finite collection, e.g. the reputation of a service may be evaluated by {*Low, veryLow, Medium, very High, High*};
- regional metric, denoted by a numerical range $[min, max]$;
- graded metric, e.g. ranking of a hotel service by $\{1, 2, 3, 4, 5\}$;
- Boolean values, numeric and enumerative scales.

It is worth to note that this QoS model is easy to extend, and that the users may customise their $\mathcal{Q}_N, \mathcal{Q}_D, \mathcal{Q}_I$ at will. Detailed definitions of all quality properties are out of our scope. Instead, we focus on how to take this QoS model as foundation to perform a combined evaluation of service qualities.

7.4.3 Selection Algorithm

QoS-based selection of services is not easy, not only due to the diversity of multifarious quality metrics with different value types, value range and measurements, but also since an effective algorithm is missing, which is supposed to evaluate all metrics in combination.

We assume that $\mathcal{Q}_R = \{r_1, r_2, ..., r_k\}$ expresses a user's quality requirements comprising k quality properties. Similarly, we assume that there are m quality profiles of the candidate web services form $resutlSet_1$, denoted as $\mathcal{Q}_S = \{\mathcal{Q}_{A_1}, \mathcal{Q}_{A_2}, ..., \mathcal{Q}_{A_m}\}$, where $\mathcal{Q}_{A_i} = \{q_{i1}, q_{i2}, ..., q_{ij}\}$, $i, j \in \mathcal{N}$. It implies that the advertisement of service s_i provides j quality metrics. As already known, there are two cases at this matchmaking stage, namely,

1. $\mathcal{Q}_R = \emptyset$, then $\mathcal{Q}_R := \mathcal{Q}_D$; and
2. $\mathcal{Q}_R \neq \emptyset$, then $\mathcal{Q}_R := \mathcal{Q}_R \cup \mathcal{Q}_N$ and $\mathcal{Q}_R$ is matched with each Q_{Ai}, $i \in \mathcal{N}$.

Quite obviously, it is rather unlikely that any $\mathcal{Q}_R$ and $\mathcal{Q}_{A_i}$ will happen to have the same number of quality metrics. So, in the first preprocessing step, we take $\mathcal{Q}_R$ as benchmark for aligning each $\mathcal{Q}_{A_i}$. This process includes,

1. to re-arrange the metrics of $\mathcal{Q}_{A_i}$ in the same order as $\mathcal{Q}_R$;
2. if $\mathcal{Q}_{A_i}$ lacks a quality property of $\mathcal{Q}_R$, then the property can be added to $\mathcal{Q}_{A_i}$ with its value set to 0; and
3. to remove the qualities of $\mathcal{Q}_{A_i}$, which are not listed in $\mathcal{Q}_R$.

Therefore, the QoS matrix for service matchmaking, $M_{\mathcal{Q}} = \{\mathcal{Q}_R, \mathcal{Q}_{A_1}, \mathcal{Q}_{A_2}, ..., \mathcal{Q}_{A_m}\}$, is presented as,

$$
M_{\mathcal{Q}} = \begin{pmatrix}
r_1 & r_2 & r_3 & \cdots & r_k \\
q_{11} & q_{12} & q_{13} & \cdots & q_{1k} \\
q_{21} & q_{22} & q_{23} & \cdots & q_{2k} \\
\cdots & \cdots & \cdots & \cdots & \cdots \\
q_{m1} & q_{m2} & q_{m3} & \cdots & q_{mk}
\end{pmatrix}_{(m+1) \times k}
\tag{7.2}
$$

$M_{\mathcal{Q}}$ is an $(m+1) \times k$ matrix, with a quality requirement $\mathcal{Q}_R$ as its first row, and the quality profiles of candidates services sequentially as the other rows. Each column contains values of the same quality property. For uniformity, matrix $M_{\mathcal{Q}}$ has to be normalised with the objective of mapping all real values to a relatively consistent range, i.e. the elements of the final matrix are real numbers in the closed interval $[0, 1]$. The main idea of this normalisation is to scale the value ranges with the maximum and minimum values of each quality metric for a number of current candidate services. Accordingly, the maximum and minimum values are mapped to the uniform values 1 and 0, respectively, which is totally depending on their definition of *hasTendency*.

For instance, a user searches for a flight and constrains the ticket price to be below € 300, and three service providers ask for € 250, € 280 and € 260, respectively. In this case the minimum and maximum are € 250 and € 280. Then, the calculations of the relative closeness of this quality metric read as $(1 - \frac{250-250}{280-250}) = 1$, $(1 - \frac{280-250}{280-250}) = 0$ and $(1 - \frac{260-250}{280-250}) = 0.667$.

The second preprocessing step is to perform a uniformity analysis. We distinguish three cases based on different quality metrics together with their value properties. By taking the value of *hasTendency* of a quality property r_i, $1 \leqslant i \leqslant k$ (cp. Section 7.4.1), we specify that,

1. if *hasTendency* $='$ *given'*, then we calculate the ratio by

$$q'_{ij} = \begin{cases} 1 - \frac{q_{max} - q_{ij}}{q_{max} - q_{min}} & \text{if } r_j \geqslant q_{max} \\ \frac{q_{ij} - q_{min}}{q_{max} - q_{min}} & \text{if } r_j \leqslant q_{min} \\ 1 - \left| \frac{|q_{ij} - r_j| - m}{n - m} \right| & \text{if } r_j \in (q_{min}, q_{max}) \end{cases} \tag{7.3}$$

2. if *hasTendency* $='$ *small/low'*, then the ratio is calculated by

$$q'_{ij} = 1 - \frac{q_{ij} - q_{min}}{q_{max} - q_{min}} \tag{7.4}$$

3. if *hasTendency* $='$ *large/high'*, then the ratio is calculated by

$$q'_{ij} = 1 - \frac{q_{max} - q_{ij}}{q_{max} - q_{min}} \tag{7.5}$$

where $1 \leqslant i \leqslant k$, $1 \leqslant j \leqslant m$ and

$$q_{max} = \max\{q_{ij}\} \tag{7.6}$$
$$q_{min} = \min\{q_{ij}\} \tag{7.7}$$
$$n = \max\{|q_{ij} - r_{ij}|\} \tag{7.8}$$
$$m = \min\{|q_{ij} - r_{ij}|\} \tag{7.9}$$

According to the algorithms defined by Formulae (7.3), (7.4) and (7.5), Fig. 7.3 illustrates these three cases of matchmaking. The course of the scale axis from the left to the right corresponds to growing values, whose tendencies are *small/low*, *given* and *large/high*, respectively. Then, the value of r_j, $1 \leqslant j \leqslant k$ is scattered either among q_{ij}, $1 \leqslant i \leqslant m$ on the right side or the left side of the candidates values.

In detail, Formula (7.3) describes the case that a user requires the value of a quality to be as close to his/her given value as possible. We assume that r_j has a value, u_j, and the other qualities $\{q_a, q_b, ..., q_h\}$ have values, $\{v_a, v_b, ..., v_h\}$. There are three cases corresponding to Formula (7.3). First, when $r_j \geqslant q_{max}$, the candidate set will be $\{q_a, q_b, q_d, q_d\}$, and by Formula (7.3) we know that q_d assumes the best ratio with 1.0. A similar situation occurs when $r_j \leqslant q_{min}$. When r_j scatters in $\{q_c, q_d, q_e, q_f\}$,

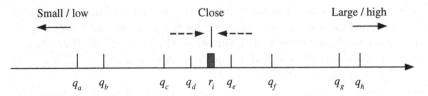

Fig. 7.3 Quality measurement

the range of the scale should first be defined by $(n - m)$, and then ratios are calculated following the third case of Formula (7.3). Analogously, Formulae (7.4) and (7.5) define the calculation methods to be employed when the tendency values of $\mathcal{Q}_R$ are *small* and *large*, respectively.

The weighted value of each quality metric is defined by the parameter *hasWeight* in Table 7.1. These are brought into the form of a diagonal matrix, $W = \{w_1, w_1, ..., w_k\}$. Here, we assume that $\sum_{i=1}^{n} w_i = 1$. Then, W is applied to matrix $M_{\mathcal{Q}}$ and yields,

$$M_{\mathcal{Q}'} = M_{\mathcal{Q}} \times W = \sum_{i=1}^{m} (q'_{ij} \cdot w_i) \tag{7.10}$$

Finally, we can calculate the evaluation result for each quality metric by summing the values of each row. These abstract values are taken as relative evaluation of each service's qualities.

7.5 Experiments and Evaluation

For comparison purposes, the test data from [93] were employed. In their experiments, the authors implemented a hypothetical telephone service registry, which provides various telephone services such as long distance, local, wireless and broadband. They simulated 600 users to collect experimental data. For example, two telephone services' test data are presented with seven quality criteria, namely *Price*, *Transaction*, *Time Out*, *Compensation Rate*, *Penalty Rate*, *Execution Duration* and *Reputation*. Their corresponding value types are $, 0/1, microseconds, percent, percent, microseconds and a rank value in $[0, 5]$.

Table 7.2 Experimental data

Data	Price	Trans-action	Time Out	Compensation Rate	Penalty Rate	Execution Duration	Reputation
R	30	1	80	0.4	0.8	120	4.0
ABC	25	1	60	0.5	0.5	100	2.0
BTT	40	1	200	0.8	0.1	40	2.5
A_1	28	1	140	0.2	0.8	200	3.0
A_2	55	1	180	0.6	0.4	170	4.0

In order to evaluate our QoS selection model, we introduce a customer's quality requirement and another two services to the experiment, which gives rise to the QoS matrix $M_{\mathcal{Q}}$ in Table 7.2. Its first row is the supposed Q_R, the next two rows are taken from [93], and the last two ones are also hypothetical candidate services. From the listed quality properties of this example, we know that *Price* and *Execution Duration* are expected to be smaller, *Compensation Rate*, *Penalty Rate* and *Reputation* are to be bigger, and *Time Out* is required to be as close to the user's expectations as possible. The result of normalisation carried out by our algorithm finally reads,

$$
Q' = \begin{pmatrix}
1 & 1 & 0.870 & 0.500 & 0.571 & 0.625 & 0 \\
0.500 & 1 & 1 & 1 & 0 & 1 & 0.250 \\
0.900 & 1 & 0.522 & 0 & 1 & 0 & 0.500 \\
0 & 1 & 0 & 0.667 & 0.429 & 0.188 & 1
\end{pmatrix}
\tag{7.11}
$$

Assuming weights $W = \{0.4, 0.0, 0.0, 0.2, 0.1, 0.1, 0.2\}$, we apply Formula (7.10) to obtain a quality evaluation set, named $Q'' = \{0.6196, 0.5500, 0.5600, 0.3951\}$. That is, in case of putting a high weight on price, service *ABC* is the best choice, the order of the results is in line with human intuition, and the result is consistent with [93] as well.

Discussion. Our QoS-based selection model considers the dynamic and real-time information, thus it is fully adapted to current distributed network environments. When kept relativity well and up to date, it is also rather fair, because it is based on comparing the currently available web services against a given service requirement. If web services are added or deleted, the evaluation should be updated accordingly.

7.6 Summary

In this chapter we presented our fourth contribution to the selection of semantic web services. A novel QoS-based approach to select semantic services was proposed by presenting a fair and simple algorithm to evaluate multiple quality metrics in combination. First, we specified a QoS ontology and its vocabulary in order to augment the QoS information in WSMO. Furthermore, various quality attributes, their respective measurements, and a QoS selection model were defined in detail. Finally, a fair and dynamic selection mechanism was presented, which uses a normalisation algorithm oriented at optimum value ranges. This approach was validated and evaluated by a case study on a set of telephone services.

Chapter 8
Evaluation

In the previous four chapters we have presented step by step our ontology- and quality-based approach to discover and select semantic web services. This chapter commences with an integrated evaluation of them. Since the main part of this work is based on ontology technology to support semantic web service applications, this evaluation involves two different contexts, ontologies and SWSs. Accordingly, it is divided into two parts and performed with the fields' respective assessment standards and methods.

As building application ontologies according to Chapter 5 is more related to ontology merging and building, for its evaluation it is better to consider the criteria usually employed by ontology applications. Since we have presented a semantic service model with a set of matchmaking algorithms facilitating service discovery, the evaluation of this part focuses on assessing its efficiency and retrieval accuracy.

Moreover, the experimental platform built in the WSMX environment (cp. Section 4.3) is used to perform various assessment exercises. Before the evaluation work is presented in detail, the generally used criteria and data sets are introduced.

8.1 Evaluation Criteria

The two common criteria *precision* and *recall* from the field of Information Retrieval (IR) can be extended and shared by both evaluation parts. Besides, comparing with golden standards or the work of experts is also a commonly used assessment method. In the following sections, we briefly present an introduction of these criteria.

8.1.1 Precision and Recall

Precision and recall are two widely used criteria to evaluate the quality of results in the domain of information retrieval [102, 181]. Precision is seen as a measure of exactness or fidelity, whereas recall is a measure of completeness. Here, they are

X. Wang and W.A. Halang: Discovery and Selection of Semantic Web Services, SCI 453, pp. 109–125.
springerlink.com   © Springer-Verlag Berlin Heidelberg 2013

especially borrowed to evaluate the results of discovering semantic web services. For related work refer to [177, 44].

In the context of semantic web services, *precision* is specifically defined as the number of relevant web services retrieved by a search divided by the total number of web services retrieved by that search (cp. Equation (8.1)), and *recall* is defined as the number of relevant web services retrieved by a search divided by the total number of existing relevant web services (which should have been retrieved; cp. Equation (8.2)),

$$Precision = \frac{|re \cap rt|}{|rt|} \tag{8.1}$$

$$Recall = \frac{|re \cap rt|}{|re|} \tag{8.2}$$

where $|re|$ is the set of **relevant** web services, which consists of all semantic services relevant to the service requirements, $|rt|$ is the set of **retrieved** web services, which includes all web services finally produced by a search engine based on a service query, and $|re \cap rt|$ is the set of relevant web services retrieved by the search engine. Obviously, the values of precision and recall are between 0 and 1. A perfect precision score of 1 means that every result retrieved by a search was relevant, whereas a perfect recall score of 1 means that all relevant web services were retrieved by the search. In conclusion, a good search engine should seek high precision and high recall.

Besides, there is an inverse relationship between *precision* and *recall*, when it is possible to increase one at the cost of reducing the other. For example, to increase the recall by retrieving more web services has the cost of increasing the number of irrelevant web services retrieved (decreasing precision). To balance them, we consider the traditional *F-measure* defined in Equation (8.3), which is an equally weighted harmonic mean of precision and recall,

$$F = 2 \cdot \frac{Precision \times Recall}{Precision + Recall} \tag{8.3}$$

In consequence, precision and recall of our service searching engine will be evaluated in the space of semantic web services collected.

8.1.2 Golden Standards

Other convincing evaluation methods are to compare results with golden standards of their respective fields or with manual work performed by domain experts. Here, a golden standard normally means a benchmark or a measurement which is widely accepted as being the best one available in a domain. For ontology merging, e.g. our merged ontology is compared to a golden standard ontology, which worked as reference ontology (cp. Section 8.3).

Obviously, the approach of golden standard evaluation assumes that it contains all correct results. In reality, though, a golden standard easily omits many potential facts and introduces new ones from other sources (such as the domain knowledge of experts). Thus, evaluation results are very likely to be influenced by these imperfections of golden standards. To compensate for these errors, expert evaluation can sometimes be performed, and ideally such an evaluation should be performed by several experts.

8.2   Experimental Data

Before presenting our evaluation results, it is necessary to introduce all experimental data sets to be used. The data sets were collected from different sources, some defined in OWL and WSDL and others in WSML languages. Thus, some preparation, e.g. data translations, is proposed in this section as well.

8.2.1   Collection of Data Sets

In order to evaluate our results in the best possible way, we have collected as many data as possible. All these datasets can directly be downloaded[1]. Based on the different sources, we currently have the following four data sets.

Data set 1: Shipment web services.   The shipment scenario of the Semantic Web Service Challenge[2] is one of our use cases for performance analysis. The SWS Challenge is a widely recognised initiative to demonstrate and compare semantically enabled web service discovery techniques based on real-world services.

   The original description of the shipment scenario (in natural language, cp. [130]) defines five web services for package shipment from the USA to different destinations along with several examples of concrete client requests (as goal instances) for web service discovery. Table 8.1 gives an overview of two provided functionalities of five web services, i.e. the conditions on locality and maximum weights. Further, the actual price for a particular shipment is defined based on the dependency between package weight and constraints on delivery time.

   Therefore, service discovery can be based on these two functional descriptions to find the functionally usable web services, and use the prices to choose the best candidate. Accordingly, a user's query which is formalised into a goal is concentrated into constraints on locality, package weight and price.

   For this scenario we have collected two separate domain ontologies, which define the relevant background knowledge. The *location ontology* essentially provides a knowledge base of all continents, countries and cities in the world

---

[1] http://www.wsao.ie
[2] http://sws-challenge.org/

Table 8.1 Overview of shipment web services

Web service	Sender	Supported destinations	Maximum weight (kg)
Muller	USA	Africa, Europe, North America, Asia	50
Racer	USA	Africa, Europe, North America, South America, Asia, Ocenania	70
Runner	USA	Europe, South America, Asia, Ocenania	heavy
Walker	USA	Africa, Europe, North America, South America, Asia, Ocenania	50
Weasel	USA	USA	heavy

(e.g. geographic relationships are defined via a transitive predicate *locatedIn* $(?x, ?y)$). The *shipment ontology* specifies the concepts package, sender, receiver, shipment order and different weight classes whose formal relationship is defined by a transitive inclusion predicate, e.g. *includeIn(light, heavy)*.

The data set includes these five web services' descriptions, two ontologies and ten goal templates (which vary on either location or weight) specified in WSML-full[3].

Data set 2: Semantic government web services.   Our methods were tested within the project SemanticGov[4] funded by the European Union, which is based on real show cases as described in [16]. The project aims to build a semantic government system employing emerging semantic web service technologies. Governmental public administration (PA) is modeled as WSMO-PA web services [174], further providing various applications (e.g. discovering PA services) to citizens at the national and pan-European levels.

The show case used here describes changes of residency in Europe, e.g. a Belgian citizen applies to move to Italy, say Turin (which is listed in Table B.2 of Appendix B). In this data set, we have collected six PA web services together with their respective Governance Enterprise Architecture (GEA) ontologies describing properties such as service type, domain or category. An example of a GEA web service ontology is listed in Table B.3 of Appendix B. These six PA web services have a similar function of changing residency with different constraints on citizenship or residential status (e.g. age, single/married). The data set contains quite complete service descriptions of non-functional and functional properties, and all are well defined in WSML-DL by domain experts. Using this data set, the θ-subsumption algorithm of Section 4.2.2.5 was thoroughly tested.

Data set 3: Travel web service ontologies.   To test the performance of our tool *ontoMerge* (cp. Fig. 5.8), we took a set of travel-related ontologies, which are mostly compiled from the web, such as the DAML Ontology Library[5] and the ontologies of Protégé[6]. This data set contains 23 travel-related ontologies, four of which are particularly used for detailed analysis and comparison with results

---

[3] http://www.wsmo.org/TR/d16/d16.1/v0.21/

[4] http://www.semantic-gov.org/

[5] http://www.daml.org/ontologies/

[6] http://protege.stanford.edu/download/ontologies.html

of other tools, e.g. the Protégé merging plug-in. Moreover, two of them[7] are listed in Tables B.4 and B.6 of Appendix B, which are the ones investigated to instruct the ontology merging tool (cp. Section 5.3.3.3).

Data set 4: Zip web services.   To be comparable, we also use the same data set as in [44, 81, 84, 94, 176]. It contains 17 zip-related web services, some of which were already presented in Section 6.1.

In fact, the fourth data set is just a part of the full data set which we have collected from the *Semantic Web Services Repository* of University College Dublin[8] and XMethods[9]. The full data set contains the raw data of 799 WSDL web services, 366 of which are not classified. The remaining ones were roughly divided by the editors into 26 major categories and 70 subcategories. The *Business* category, for instance, has 23 web services and 3 subcategories, viz. *adverts, charts* and *customers*.

8.2.2 Data Preprocessing

Since part of the data collected on web services are web service descriptions either in WSDL or OWL, but our approaches are implemented in WSMO/WSML, some data preprocessing to generate WSML web services and service ontologies from the WSDL or OWL service descriptions is necessary.

Here, preprocessing was performed for data set 4, only. In detail, the zip web services of data set 4 are initially modeled by OWL-S services together with their *WSDL* descriptions. Usually any web service consists of five documents, which are *_service.owl, _profile.owl, _process.owl, _Grounding.owl* and *_Concepts.owl*. For example, in Appendix B, Tables B.7, B.8 and B.9 list the related descriptions of the web service *zip-code-Lookup*. These untreated data are first transferred from OWL-DL to WSML-DL format. The translation rules are specified in Table 8.2 by referring to [155]. Finally, 17 valid *zip*-related web services are obtained in WSML-DL language. During translation, the following problems are encountered.

1. The semantic information provided by service *_Concepts.owl* is not well modeled with respect to OWL syntax which, in turn, limits the possibilities of service matchmaking. In particular, definitions of many concepts are missing and some concepts were not well defined, e.g. a concept *"city"* is defined as *concept* and *relation* at the same time.

2. A very limited number of constructs is actually used by every *_Concepts.owl*, thus quite poor service ontology information is provided without proper concept definitions and constraints. All constructs were used, including *owl:class, owl:ObjectProperty, owl:DatatypeProperty, rdfs:range, rdfs:domain, rdfs:sub, ClassOf* and *rdfs:subPropertyOf*, which are, respectively, translated to *concept,*

---

[7] Taken from `http://deri.org/iswc2005tutorial/ontologies/`

[8] `http://moguntia.ucd.ie/repository/`

[9] `http://www.xmethods.net/ve2/Directory.po`

Table 8.2 Mappings of OWL-DL and WSML-DL

OWL-DL	WSML-DL
owl:class	**concept**
rdf:datatype	**datatype**
rdf:Property	**attribute**
owl:ObjectProperty(prop_id **domain** (id1) **range**(id2))	**concept** id1 prop_id **impliesType** id2
owl:DatatypeProperty(prop_id **domain** (id) **range**(xsd:Datatype))	**concept** id prop_id **ofType** xsd#Datatype
rdfs:subPropertyOf(prop_id1,prop_id2)	?x[prop_id2 **hasValue** ?y] **impliesBy** ?x[prop_id1 **hasValue** ?y]
owl:restriction(prop_id **allValuesFrom** ELEMENT)	**forall** ?x(?y[prop_id **hasValue** ?x] **impliesType** ELEMENT)
owl:restriction(prop_id **someValuesFrom** ELEMENT)	**exists** ?x(?y[prop_id **hasValue** ?x] **and** ELEMENT)
owl:equivalentClass	synonym (**relation**)
owl:sameAs	synonym (**relation**)
rdf:subClassOf	specialisation (**relation**)
owl:individuals	**instances**

subConceptOf, *attribute* and *datatype* in WSML-DL. An example is shown in Table 8.3.

The modeling deficiencies of practical service ontologies increase the difficulties of evaluation. This problem is most likely caused by the fact that semantic services are created by service providers who are usually not experts of domain knowledge and lack ontology engineering skills. These problems also reveal the fact that for any kind of semantic web services a uniform application ontology is needed.

Based on Definition 5.1 the initial service ontologies are produced after the pre-processing above. Taking the service description of 115_Zip_Code_Lookup_Service as example, the service ontology generated is visualised in Fig. 8.1 and listed in Table B.8 of Appendix B. It contains four concepts and two attributes with the _string datatype.

Summary. So far, we have described in detail four data sets which are used to evaluate our work. These data sets are all real ones related to different kinds of web services. An overview of them is presented in Table 8.4.

From Table 8.4 we know that for the *Shipment* and *SemanticGov* web services respective application ontologies are defined. They do not need to consider ontology building and are mainly used to test our function-based service discovery mechanism. No ontologies are defined for the *Travel* web services. They are primarily used to test our ontology merging and building approaches. The *Zip*-related web services' data set contains service descriptions and their service ontologies, which are used to assess the ontology-based web service selection method. As all these data sets do not belong to the same web service category, they provide a realistic test environment to evaluate the feasibility of our service discovery approach.

Table 8.3 Translation example

OWL-Lite Example	Translated to WSML-DL

`<owl:Class rdf:ID="CityStateToZipCodeResponse_45582">`
`<rdfs:subClassOf rdf:resource=`
`"http://moguntia.ucd.ie/daml/Datatypes.daml#ZIP_Code"/>`
`</owl:Class>`
	concept CityStateToZipCodeResponse
	    subConceptOf ZIP_Code
	concept ZIP_Code
	instance
	CityStateToZipCodeResponse_45582

`<owl:ObjectProperty rdf:ID="CityStateToZipCodeResult_45595">`
`<rdfs:range rdf:resource="#ArrayOfString_45592"/>`
`<rdfs:domain`
`rdf:resource="#CityStateToZipCodeResponse_45582"/>`
`<rdfs:subPropertyOf rdf:resource=`
`"http://moguntia.ucd.ie/daml/Datatypes.daml#ZIP_Code"/>`
`</owl:ObjectProperty>`
	concept CityStateToZipCodeResponse
	    CityStateToZipCodeResult
	    impliesTypes ArrayOfString
	
	axiom
	  definedBy
	    ?x[ZIP_Code **hasValue** ?y]
	  impliedBy ?x[CityStateTo
	  ZipCodeResult **hasValue** ?y]
	
	instance
	    CityStateToZipCodeResult_45595
	instance ArrayOfString_45592

`<owl:Class rdf:ID="CityToZipCode_45583">`
`<rdfs:subClassOf rdf:resource=`
`"http://moguntia.ucd.ie/daml/Datatypes.daml#City_Name"/>`
`</owl:Class>`
	concept CityToZipCode
	    subConceptOf City_Name
	concept City_Name
	
	instance CityToZipCode_45583

`<owl:DatatypeProperty rdf:ID="City_45596">`
`<rdfs:range`
`rdf:resource="http://www.w3.org/2001/XMLSchema#string"/>`
`<rdfs:domain rdf:resource="#CityToZipCode_45583"/>`
`<rdfs:subPropertyOf rdf:resource=`
`"http://moguntia.ucd.ie/daml/Datatypes.daml#City_Name"/>`
`</owl:DatatypeProperty>`
	concept CityToZipCode
	    city **ofType** xsd#string
	
	instance City_455963

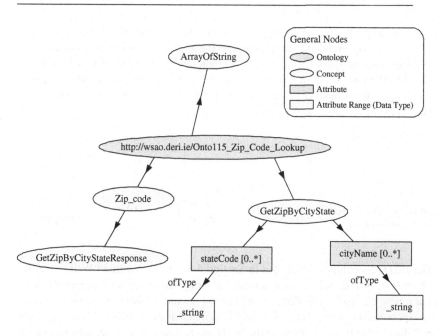

Fig. 8.1 Ontology of 115_Zip_Code_Lookup_Service

Table 8.4 Overview of test data sets

Data sets	Web services	Ontologies	pre-Goals	Language
Shipment	5	3	10	WSML
SemanticGov	6	2	–	WSML
Travel	2	23	–	WSML & OWL
Zip	17	17	–	OWL & WSDL
full set (raw)	799	164	–	OWL & WSDL

8.3   Application Ontology Evaluation

As mentioned in Section 5.4, supported by the EU projects SemanticGov (FP6-207527) and KnowledgeWeb (FP6-507482) we have developed a scalable experimental platform, *WSAO* Studio, as reference implementation of our service selection architecture presented in Fig. 6.1. It is an Eclipse 3.0 plug-in composing a set of tools, such as *WSD* of Fig. 5.7, *OntoMerge* of Fig. 5.8 and *OntoBuilder* of Fig. 5.9. In the following sections, the related evaluation of these subtools is presented.

8.3.1   Performance of Word Sense Disambiguation

As the first step of building an application ontology, word sense disambiguation (WSD) is very important, and directly affects the quality of the application ontology built. WSD is commonly recognised as a challenge in the field of ontology engineering [5, 9]. Here, we do not only address this problem, but for the first time it is clearly proposed to consider word sense disambiguation before merging ontologies in the context of semantic web services.

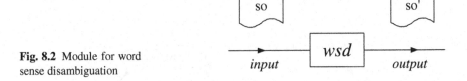

Fig. 8.2 Module for word sense disambiguation

The *WSD* algorithm was detailed in Section 5.3.2.1. A simplified WSD module, as illustrated by Fig. 8.2, takes a service ontology (*so*) as input and generates as output a new one (*so'*). The difference between *so* and *so'* consists of two aspects: (1) *so'* specifies the right senses of its words as much as possible and (2) *so'* contains more information than *so*, because the *WSD* module has an information-retrieving function able to extract the related information from WordNet based on the words' right senses (cp. Section 5.3.2.2). Hence, there are two ways to evaluate the work of *WSD*:

1. How many word senses have correctly been disambiguated? Assume that the initial *so* has n polysemous concepts, and for m $(0 \leqslant m \leqslant n)$ of them the right senses were found by performing the WSD algorithm, then the precision, Pre_{wsd}, is defined as

$$Pre_{wsd} = \frac{m}{n} \in [0, 1] \tag{8.4}$$

Obviously, the more right word senses are found, the better is the result.

2. How much information is added? The more related information is extracted, the more semantics is available. Assume that the *information content* is expressed as sum u of the number of concepts, attributes, instances and concept relations of *so*, and correspondingly u' of *so'*, then the information augmentation Aug_{wsd} is defined as

$$Aug_{wsd} = \frac{u' - u}{u} \in [0, 1] \tag{8.5}$$

The *travel*1 ontology (listed in Table B.6 of Appendix B) has, for instance, initially 11 concepts (including one composite concept *trainTimeTable*), 39 attributes, 3 instances and 0 relations. After parsing these words, we obtained 12 formal concepts. Then, we executed the *WSD* operation and found that 11 concepts are polysemous, 2 of them failed to find their right senses. Two subrelations between *date* and *time* and between *customer* and *person* were newly discovered. Besides, based on the right word senses we found 7 synonyms, e.g. *customer#1*'s synonym is *client* and *cost#2*'s synonyms are *monetary value* and *price*, but we only extracted one pair of synonyms, *cost* and *price*, based on our *trade-off principle*. In consequence, for ontology *travel*1, we have $Pre_{wsd} = \frac{11-2}{11} = 0.8181$ and $Aug_{wsd} = \frac{11+39+3+2+1}{11+39+3+0} - 1 = 0.0567$. Analogously, another four examples are listed in Fig. 8.5.

Table 8.5 Some results of *WSD* evaluation

Test onto.	Concepts	Initial info.	New Info.	Pre_{wsd}	Aug_{wsd}
#1	11	53	56	81.82%	5.67%
#2	11	36	39	100%	8.33%
#3	16	42	42	93.33%	0.00%
#4	9	36	37	77.78%	2.78%
#5	50	74	79	88.89%	6.76%
Average	–	–	–	88.34%	4.71%

In total, we collected about 50 results on experimental ontologies. Their precision of word sense disambiguation is very good with an average of 88.23%. Also the information augmentation percentage of 3.52% is quite good. In fact, we do not expect to mine too much information, as it will increase computational complexity. According to the *trade-off principle*, if mined information involves two concepts of its original ontology, then this information is added to the result ontology. Such a principle is used to keep information augmentation at a relatively good level.

Discussion. A lesson learnt from these experiments is that a good knowledge base, e.g. a dictionary, is the key factor to success. Although WordNet is a well structured, has a large lexical database and is useful for designing application programming interfaces, it is still weak in its knowledge organisation, i.e. the relations between concepts need refinement, and thus in supporting ontology building.

8.3.2   Performance of Merging Service Ontologies

Our approach of semantic service discovery is actually based on an application ontology built by merging the same type of service ontologies. Therefore, after *WSD*, we considered to merge service ontologies two by two. The *OntoMerge* prototype was presented in Section 5.3.2.1. Simplified in Fig. 8.3, the ontology merging module takes two service ontologies, so'_1 and so'_2, as inputs and the output is a new ontology named so''.

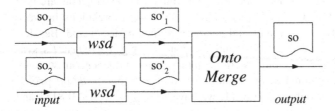

Fig. 8.3 Module to merge ontologies

The evaluation of ontology merging, building or learning is not straightforward, however, because there are no standard benchmarks available nor well defined criteria for them. How to evaluate a merged ontology is what we should be concerned with here. The best possible way for that is to compare our result ontology (O_m) with other people's work (O_p) or with a predefined golden standard (O_g). When it comes to frequent and large-scale evaluations and comparisons of multiple ontology learning approaches, according to [39] only comparison with golden standards is feasible in practice.

To compare two ontologies [99], the *lexical* and *conceptual* levels need to be considered. Lexical comparison assesses the similarity between the lexica (set of labels denoting concepts) of the two ontologies. At the conceptual level, the taxonomic structures and the relations in the ontologies are compared. These comparison methods as specified in detail by [39] are used here for our assessments.

Given a computed ontology O_C and a reference ontology O_R, the lexical precision (LP) and lexical recall (LR) are defined as,

$$LP(\mathcal{O}_C, \mathcal{O}_R) = \frac{|C_C \cap C_R|}{|C_C|} \qquad LR(\mathcal{O}_C, \mathcal{O}_R) = \frac{|C_C \cap C_R|}{|C_R|} \qquad (8.6)$$

Lexical precision and recall reflect how good the lexical terms computed cover the target domain. At the conceptual level, we are then concerned with the concept hierarchy or taxonomy overlap (tp, defined by [99]), which is suggested to characterise a concept by its semantic cotopy, i.e. all its super- and subconcepts. Given a concept $c \in C$ and the ontology $\mathcal{O}$, the semantic cotopy sc is defined as,

$$sc(c, \mathcal{O}) := \{c_i | c_i \in C \wedge (c_i \leqslant c \vee c \leqslant c_i)\} \qquad (8.7)$$

$$tp_{sc}(c_1, c_2, \mathcal{O}_C, \mathcal{O}_R) := \frac{|sc(c_1, O_C) \cap sc(c_2, O_R)|}{|sc(c_1, O_C)|} \qquad (8.8)$$

Further, the global taxonomic precision and recall are defined based on the semantic concept cotopy and its *local taxonomic precision* (as Equation (8.8)) and *recall* as,

$$TP_{sc}(O_C, O_R) := \frac{1}{|C_C|} \sum_{c \in C_C} \begin{cases} tp_{sc}(c, c, O_C, O_R) \text{, if } c \in C_R \\ 0 \qquad \text{, otherwise} \end{cases} \qquad (8.9)$$

$$TR_{sc}(O_C, O_R) := TP_{sc}(O_R, O_C) \qquad (8.10)$$

Since precision and recall are inverse to each other, we have Equation (8.10). All in all, the measures TP_{sc} and TR_{sc} do not allow a separate evaluation of lexical term layer and concept hierarchy.

We assume that the computed ontologies are our merged one, O_m, and the result generated by the Protégé plug-in for managing multiple ontologies PROMPT[10], O_p. The merged ontology prepared by domain experts (five domain experts were invited for these two experiments), is called reference ontology, O_r. Therefore, we compare O_m and O_p with O_r, respectively.

In the following, two groups and four ontologies are taken for this part of the evaluation. The first group has the ontologies *airReservation.owl* and *carRental.owl*, and the second one the two commonly used ontologies taken from the ISWC05 tutorial[11]. Employing Equations (8.6), (8.9) and (8.10), the evaluation results are collected in Tables 8.6 and 8.7.

Table 8.6 Evaluation for the ontologies of *Group 1*

Compare O_r with	LP	LR	TP_{sc}	TR_{sc}	TF_{sc}
O_m	100%	100%	98.8%	100%	99.4%
O_p	100%	100%	88.1%	100%	93.6%

---

[10] http://protege.cim3.net/cgi-bin/wiki.pl?Prompt
[11] http://deri.org/iswc2005tutorial/ontologies/

Table 8.7 Evaluation for the ontologies of *Group 2*

Compare O_r with	LP	LR	TP_{sc}	TR_{sc}	TF_{sc}
O_m	100%	100%	97.2%	92%	94.5%
O_p	100%	96%	89.5%	86%	87.7%

Discussion. Looking at the results presented in Tables 8.6 and 8.7, one can see that our merging result is better. The reasons are that (1) our merging approach preserved the information of the merged concepts by creating a synonym concept relation resulting in precision and recall at the lexical level to be both 100%, and that (2) our merging approach is based on WordNet to mine any potential relations between concepts, rendering the merged ontology to be more reasonable and closer to the reference ontology with higher value of TP_{sc}.

During the process of ontology merging, the semantical similarity of ontological concepts was calculated based on our concept similarity algorithm defined by Equation (6.4). The concept similarity results of Group 1 are listed in Table 8.8. Such ontological concept similarity is finally used to measure semantic web service similarity.

8.4 Evaluation of Semantic Service Similarity

Finally, we commence with the evaluation of our semantic web service similarity techniques. In Chapter 4, we have specified a quadruple service model for selection, and we proposed its respective similarity algorithms. In this section these algorithms and their contribution to web service discovery and selection are to be evaluated in an integrated way. Specifically, we consider two aspects, viz. the fulfillment of functional service properties and the similarity of non-functional service properties.

8.4.1 Performance of Service Discovery

To evaluate how far services fulfill their functional requirements, we focus on testing the θ-subsumption matchmaking defined by Equation (4.23) during the process of web service discovery. We have built a query tool for PA web services and tested it in the SemanticGov project. As shown in Fig. 8.4, service queries consist of two parts, one are the logical expressions on the GEA ontology, i.e. *PADomain*, *PublicServiceType*, *Location* or *EffectType*, and the other are the PA service pre-conditions.

For example, searching for PA web services with the query listed in Table 8.9 yields two PA web services. These two PA services matched give rise to the one

Table 8.8 Semantical concept similarity of ontologies *tavel1* and *travel2*

Concepts of Onto. *Travel1*	Concepts of Onto. *Travel2*	Semantical similarity
travel1#date	travel2#date	1.0
travel1#time	travel2#time	1.0
travel1#ticket	travel2#travelVocher	0.8835978835978835
travel1#ticket	travel2#tipPoints	0.7
travel1#person	travel2#woman	0.5952380952380952
travel1#person	travel2#man	0.5952380952380952
travel1#person	travel2#human	0.5952380952380952
travel1#customer	travel2#station	0.5
travel1#customer	travel2#name	0.5
travel1#delivery	travel2#date	0.5
travel1#trainTimeTable	travel2#tripPoints	0.4
travel1#cost	travel2#payment	0.4
travel1#trainTimeTable	travel2#date	0.3333333333333333
travel1#date	travel2#travelVocher	0.3333333333333333
travel1#term	travel2#date	0.2857142857142857
travel1#ticket	travel2#name	0.2857142857142857
travel1#term	travel2#marriage	0.25
travel1#customer	travel2#human	0.2222222222222222
travel1#time	travel2#travelVoucher	0.2222222222222222
travel1#customer	travel2#man	0.2222222222222222
travel1#customer	travel2#woman	0.2222222222222222
travel1#cost	travel2#travelVoucher	0.2
travel1#terms	travel2#payment	0.2
travel1#delivery	travel2#travelVoucher	0.18181818181818182
travel1#ticket	travel2#time	0.16666666666666666
travel1#ticket	travel2#date	0.16666666666666666
travel1#person	travel2#name	0.14523809523809522
travel1#customer	travel2#travelVoucher	0.1388888888888889
travel1#customer	travel2#marriage t	0.1111111111111111
travel1#terms	travel2#travelVoucher t	0.1111111111111111
travel1#trainTimeTable	travel2#travelVoucher	0.09523809523809523
travel1#person	travel2#marriage	0.05
travel1#person	travel2#travelVoucher	0.031746031746031744
travel1#ticket	travel2#marriage	0.018518518518518517
travel1#ticket	travel2#payment	0.016666666666666666

Table 8.9 A query for PA web services in WSML DL

?x **memberOf** GEA#PublicService
and ?x[hasEffectType **hasValue** Belgium#UpdateResidenceInformation
and ?y **memberOf** Person
and ?y[hasAge **hasValue** ?age] **and** ?age > 18.

named *Turin_BelgiumChangeResidenceService.wsml* (a pre-condition of which is
age > 20) and the other has a pre-condition of *age* > 18. Obviously, these two PA
services both qualify as candidate services.

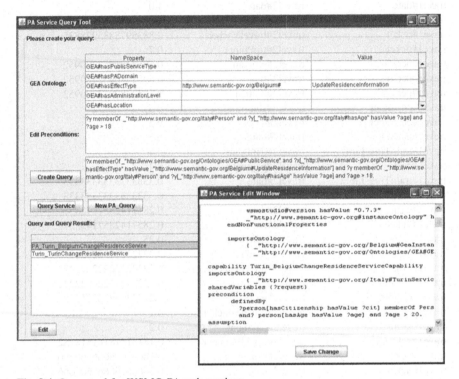

Fig. 8.4 Query tool for WSMO-PA web services

The data set 1 was used to evaluate our service discovery algorithm defined in
Section 4.3.2. Five web services of this data set, i.e. *Muller, Racer, Runner, Walker*
and *Weasel*, were tested with the 10 goal templates provided, which vary in the
constraints of shipment locations and weights. Referring to the matching degrees
defined in Section 4.3.2, the service discovery results are listed in Table 8.10. In
comparison with the results of [157] we achieved 100% precision and 100% recall.

The experiments above reveal that service discovery can find a set of candidate
web services, but to discover the service matching best a matchmaking technique
based on service similarity is required.

8.4.2 Performance of Determining Ontology-Based Service Similarity

Our main contribution is to propose an ontology-based approach for improving
web service discovery. The entire Chapter 5 addressed the building of application

Table 8.10 Results of discovering shipment services

Goal templates	**Muller**	**Racer**	**Runner**	**Walker**	**Weasel**
gtNA2NAlight	*intersect*	*intersect*	*fail*	*intersect*	*intersect*
gtRoot	*subsume*	*subsume*	*subsume*	*subsume*	*subsume*
gtUS2AF	*intersect*	*intersect*	*fail*	*intersect*	*fail*
gtUS2AS	*intersect*	*intersect*	*plugin*	*intersect*	*fail*
gtUS2EU	*intersect*	*intersect*	*plugin*	*intersect*	*fail*
gtUS2EULight	*plugin*	*plugin*	*plugin*	*plugin*	*fail*
gtUS2NA	*intersect*	*intersect*	*fail*	*intersect*	*subsume*
gtUS2OC	*fail*	*intersect*	*plugin*	*intersect*	*fail*
gtUS2SA	*fail*	*intersect*	*plugin*	*intersect*	*fail*
gtUS2World	*subsume*	*subsume*	*subsume*	*subsume*	*subsume*

ontologies. In this section we aim to evaluate how this approach can improve the retrieval accuracy in discovering semantic web services by measuring service similarity.

According to the related work mentioned in Section 3.3, keyword-based and structure-based similarity matchmaking of service descriptions are commonly used methods. Our approach, however, is based on calculated ontological concept similarity to evaluate the similarity of semantic service descriptions in order to rank the candidate web services and to discover the one matching best. Semantic service similarity was defined by Equation (6.11) and is quoted here as,

$$sim_{Service} = \sum sim_{Concept} + \sum sim_{Operation} \tag{8.11}$$

where $sim_{Concept}$ is the sum over the similarities of conceptual service information (i.e. service name, service category and operation names) and $sim_{Operation}$ is the sum over the similarities of all operations' input/output parameters with their respective data types.

Data set 4 was applied for this part of the evaluation, because it is similar to the one used by [176, 44, 84, 94, 81]. From Section 8.2.1 we know that this data set has 17 valid *zip*-related services in WSML-DL language. Table 8.11 lists its relevant statistical data. As we can see, there are more composed concepts in the initial data set, and two kinds of relationships appear in it, i.e. *Specialisation* and *Hyponym*.

In contrast, the processed data set has more formal concepts, concepts with attributes and concept relations. The reason is that our approach parsed the composite concepts as formal concepts and used WordNet to identify more concept relations, i.e. *synonym, meronym* and *hyponym*. Besides, these 17 services have 59 service operations in total with 118 input/output messages. Some examples of *zip*-related web services are listed in Table 8.12.

In order to conduct a comprehensive evaluation, we did not limit our test to the same application domain, but considered many other web services of different domains from our whole data set (described in Section 8.2.1). Finally, the experimental data set had 164 web services with 556 service operations, e.g. 17 zip-related web

Table 8.11 Statistical data of *zip* web services

Web services	Formal concepts	Composite concepts	Concepts (Attr.)	Relation types	Relations	Service names	Service operations (Msgs)
Initial data set	39	76	30 (79)	2	47	17	59 (118)
Processed data set	46	67	28 (74)	4	29	17	59 (118)

Table 8.12 Examples of *zip* web services

Service name	Service operations
zipCodeService	findZipCodeDistance
	findZipCoordinates
	findZipDetails
	getCodeSet
ZipCode	LicenseKey
	GetCityInfo
DistanceService	getLocation
	getState
	getCity
	getDistance
	getLatitude
	getLongitude
ZipCode	zipCodeToCityState
	cityToZipCode
	cityStateToZipCode
	zipCodeToAreaCode
	zipCodeToTimeZone
	cityStateToAreaCode
...	...

services, 13 weather web services, 16 stock web services etc. But only for the set of *zip*-related web services an application ontology was built.

We took each zip-related web service for matching with all other web services in order to find the similar ones. More specifically, we performed two types of experiments with our similarity algorithms: (1) name-based service similarity matching, and (2) operations-based service similarity matching. After 17 rounds of service matchmaking, we calculated the average precision and recall results by comparing them with the results elaborated by three experts of semantic web service technologies.

The comparison of our results with the average precision and recall reported by [84] and [124] is shown in Table 8.13. In these experiments we assumed that concept relations have the same weights of 0.85. Although weakening the influences of multiple relations, the same weights were used to provide feasible results and to be comparable with the precision/recall of the related work.

As shown by Table 8.13 and Fig. 8.5, our method renders much higher precision and recall than the three references. The main reasons for this are (1) that, in

Table 8.13 Results of comparing web service similarity matchmaking

Comparing to	Recall (avg.)	Precision (avg.)	F-Measure
Keyword-based [84]	19.8%	16.4 %	23.7 %
Semantic matchmaking [124]	62.6 %	63.2 %	62.89 %
Interface-based [84]	78.2%	71 %	74.43 %
Our work on service names	100 %	76.47 %	86.66 %
Our work on service operations	70.24 %	88.06 %	78.15 %

contrast to the references, our similarity evaluation is not limited to small pieces of ontologies, but uses a well structured application ontology with enriched semantics, (2) that our ontology-based service similarity is based on multiple instead of single concept relations, and (3) that our method can deal with composite concepts, which are often used in semantic web service contexts, whereas the related work avoided or was weak on this aspect.

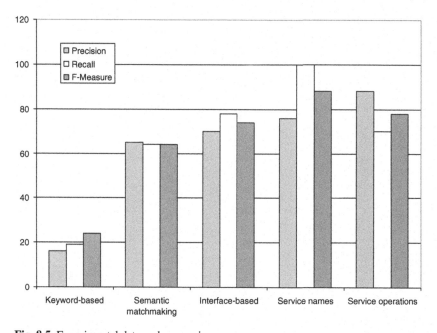

Fig. 8.5 Experimental data and comparison

8.5 Summary

This chapter presented an integrated evaluation of our approaches. After introducing the commonly used evaluation criteria, the collected data sets were detailed. Since our proposed service discovery approach mostly bases on application ontologies built, we first evaluated their quality. Then, we showed that our approach of semantic service discovery has much better retrieval accuracy by using application ontologies.

Chapter 9
Conclusion and Directions for Future Research

The goal of this book was to improve the discovery techniques of semantic web services by solving the problems and challenges remaining with the current approaches. In order to achieve this goal, the following four research questions were investigated:

1. *What kind of service model is appropriate for semantic web service discovery?*
2. *How to build an application ontology for a kind of semantic web services?*
3. *How to improve service discovery by measuring service similarity?*
4. *How to select web services by a combined evaluation of service qualities?*

After an overview of the current knowledge on the semantic web and semantic web service technologies, this book started with addressing the first research question. We defined a semantic service model and showed how it performs service matchmaking with a set of selection algorithms. In the following part, we focused on the second research question of building application ontologies, which was proposed as a novel and effective approach to mediate between heterogenous service ontologies. Then, the third and fourth research questions were addressed based on the service model defined above and the application ontologies built, aiming to enhance the quality of service discovery and selection.

This chapter is to present a detailed conclusion of this book by summarising the contributions to the four research questions above, and then to briefly outline potential directions for further research in Section 9.2.

9.1 Conclusion and Contributions

9.1.1 Service Model to Discover Semantic Web Services

By the current efforts to represent services (e.g. OWL-S or WSMO), many kinds of mechanisms (cp. Section 3.3) were considered for matchmaking web service descriptions with service requirements. The main reason of their inefficacy is that a

X. Wang and W.A. Halang: Discovery and Selection of Semantic Web Services, SCI 453, pp. 127–131.
springerlink.com   © Springer-Verlag Berlin Heidelberg 2013

specific service model is still missing which could foster selection. The approaches based on matching simple keywords, complete service profiles or even XML or RDF schemata are insufficient to describe all features necessary in service selection.

Therefore, our first contribution (cp. Chapter 4) was to define a service model apt for service discovery, being a quadruple $s = \{NF, F, Q, C\}$. This service model fully describes all characteristics necessary for semantic service discovery, namely non-functional properties (NF, further elaborated as $serviceName, serviceCategory, serviceDesciption$) and functional properties (F, elaborated as $IPOEs$), the features related to service qualities (Q) as well as an optional and extended property, i.e. the cost of service (C), with the perspective of improving the efficiency of service discovery.

The elements of the proposed service model were formally defined in Chapter 4. They fall into three categories based on their data types. The corresponding match-making algorithms are then specified as follows,

concept-based matchmaking   processes concept-related information, i.e. *service Name, serviceCategory, operationName*, and the parameters of *inputs, outputs* and service *qualities*. This kind of matchmaking is based on measuring the similarity of ontological concepts and was described in Chapter 6;

text-based matchmaking   is used to process *serviceDescription*, which normally is a short text to describe a web service. Such information is matched by the traditional *cosine/TFIDF* text similarity methods as proposed by [163];

axiom-based matchmaking   processes all constraint information on service *inputs, outputs, preconditions, effects* and on service qualities. Such kind of match-making uses a logic-based subsumption reasoning method.

These three types of matchmaking mechanisms were incorporated into two formal components of the *xSESA* architecture of WSMX.

Our contributions are to (1) specify a sound and formal service model for semantic web service discovery and to (2) define two novel matchmaking algorithms for this service model.

9.1.2 Building Application Ontologies

The fact, that heterogeneous service ontologies are used for the same kind of semantic web services, is causing many problems and impairs the interoperation between machines and web services. Much effort is required for mediating service ontologies, especially every time when communicating between services, processes and knowledge bases. Building application ontologies, $\mathscr{AO}$, for certain kinds of services is our solution to this problem.

In Chapter 5 we presented our approach to build an application ontology by a rule-based ontology merging technology. At first, we distinguished three levels of ontologies in the context of semantic web services, i.e. *generic ontology, application ontology* and *service ontology*. We presented a formal definition of ontology as a tuple of concepts, concept relations, attributes and instances in WSMO. Before building an $\mathscr{AO}$, three steps of preparatory treatment have to be considered:

1. to re-formalise the original service ontologies based on our ontology definition;
2. to perform the *WordNet-based word sense disambiguation algorithm* (*WSD*) on service ontologies; and
3. to perform the *WordNet-based concept extraction algorithm* (*WCEA*) on service ontologies in order to obtain new service ontologies with more concept relations and synonyms, and to remove concept conflicts, if any.

At last, a *WordNet-based ontology merging algorithm* (*WOMA*) is performed to merge service ontologies, in order to build an application ontology for a given application domain.

Our contributions are to propose, in the context of semantic web services, a set of algorithms (1) for initial word sense disambiguation, (2) for WordNet-based concept extraction to formalise service ontologies including the removal of conflicts between concepts and concept relations and (3) for rule-based and WordNet-based ontology merging.

9.1.3 Ontology-Based Service Selection

Although there are a few ontology-based approaches proposed for web service discovery, they are mainly based on two pieces of service ontologies, not the application domain ontology as we use, and few of them consider the full semantics expressed by multiple concept relations in the course of service matchmaking. Therefore, we presented a novel ontology-based approach by extending ontological concept relations from single-fold to manifold ones in our application ontologies (cp. Chapter 5), and represented them in a semantic net. Then, ontological concept similarity is calculated in order to assess service similarity.

In detail, four kinds of concept relations were described in Chapter 6, viz. *specialisation*, *hypernym*, *meronym* and *similarity* (including synonym). After formally defining application ontologies in WSMO, an approach to measure ontological concept similarity was proposed, which takes the shorted path between concepts and the depth of two concepts with the same, but not necessarily immediate, ancestors as metric of similarity. Based on concept similarity, we finally proposed a way to determine the similarity of semantic web services quantitatively.

Our contributions are to (1) introduce multiple concept relations into application ontologies, which are able to enhance semantical expressibility, (2) propose a novel algorithm to determine similarity of ontological concepts, both formal and composite ones, in the context of semantic web services, and (3) to propose an ontology-based algorithm to assess service similarity, which improves the accuracy of semantic service discovery.

9.1.4 QoS-Based Service Selection

The properties of service quality were thoroughly studied in the field of semantic web services. For matching service descriptions, however, the diversity of service qualities constitutes a problem. Since different service qualities have different types and require different metrics, it is hard to measure all qualities in a common way. To overcome this problem, a novel normalisation of service qualities was presented to synthetically measure service qualities during service discovery and selection.

In Chapter 7, a specific QoS ontology was defined in WSMO. Three QoS models were further distinguished, i.e. the necessary quality model, Q_N, the optional model, Q_O, and the default quality model, Q_D. For the purpose of service matchmaking, a QoS matrix $M_Q = \{Q_R, Q_{A_1}, Q_{A_2}, ..., Q_{A_m}\}$ was composed, in which $Q_R = \{r_1, r_2, ..., r_k\}$ denote the qualities expected by the service requirements, and the Q_{A_i} state the candidate services' qualities. Finally, we synthetically evaluated the metric proximity in all quality properties by a normalisation algorithm taking Q_R as benchmark. Telephone-related web services were employed to validate our approach.

Our contributions are to (1) define a detailed QoS ontology for semantic web services, (2) propose a novel normalisation algorithm for any application with the need to evaluate multiple features and (3) to devise a fair and dynamic QoS-based mechanism for service selection using a normalisation algorithm oriented at optimum value ranges in synthetical evaluation.

9.2 Directions for Future Research

Although this book focused on the discovery of semantic web services, its service model and approach to build application ontologies have resolved the core issues of semantic web services, which form the basis of many applications. Thus, actually a foundation for the success of many applications was laid. The following directions for future research may potentially benefit from this work.

- *Uncertain or fuzzy service discovery.* In general, user requirements tend to be invariably ambiguous. Even in more professional domains, where the user is ideally expected to provide exact service requirements, it is often difficult to retrieve all trivial details necessary for an exact match. Uncertain discovery means to take limited or unclear information given by the user as basis to discover services meeting his/her real expectations. With richer semantics provided by application domain ontologies and semantic reasoning technologies, it may become possible to extract, to a large extent, the implicit information in order to achieve good results in service discovery.
- *Composition of semantic web services.* As discussed in previous sections, several of our results with respect to service discovery and ontology building can be re-used when composing different services for complex business applications.

Service composition may work at the service, operation and data levels, and our application ontology approach can definitely support it at the data level. Further work on the operation level (especially state-enabled operation similarity) or on behaviour/operation flow and order could also be beneficial for service composition.

- *Service customisation and recommendation.* The future semantic web services are expected to have intelligent capabilities of customisation according to user requirements, e.g. with respect to input and output formats, in order to provide personalisation. Likely candidates may be recommended on the basis of analyses and elicitation of user interests. An application domain ontology is the best knowledge base for these tasks.

Appendix A
Description Logic, TBox and ABox

As a logic-based knowledge representation language, Description Logic (DL) is used to represent the terminological knowledge of an application domain ($\mathscr{D}$) in a structured and formally well-understood way. A DL system is characterised by four fundamental aspects [56],

- a set of **constructs**, which are used by *concept* (modeling classes of individuals) and *rôle* expressions (modeling relationships between classes) as illustrated in Table. A.1;
- a set of **TBox**, which are assertions on concepts describing a vocabulary associated with a set of facts ABox, e.g. $HappyMan \equiv Human \sqcap \neg Female \sqcap (\exists married.Doctor) \sqcap (\forall hasChild.(Doctor \sqcap Professor))$ defines a complex concept;
- a set of **ABox**, which are assertions on individuals describing facts associated with a terminological vocabulary within a knowledge base, e.g. $HappyMan$ $(Franz)$ and $child(Franz, Luisa)$; and
- the **inference mechanisms** for reasoning on both the TBox and the ABox involving to compute the subsumption relation between two concept expressions, i.e. verifying whether one expression denotes a subset of the objects denoted by another expressions.

Table A.1 DL constructors

Construct	Syntax	Semantics	Example
atomic class	C	$C^I \subseteq \Delta^I$	$Human$
atomic rôle	r	$r^I \subseteq \Delta^I \times \Delta^I$	$hasChild$
conjunction	$C_1 \sqcap C_2$	$C_1^I \sqcap C_2^I$	$Human \sqcap Male$
disjunction	$C_1 \sqcup C_2$	$C_1^I \sqcup C_2^I$	$Doctor \sqcup Lawyer$
negation	$\neg C$	$\Delta^I \perp C^I$	$\neg Male$
universal	$\forall r.C$	$\{s \mid \forall s'.(s,s') \in r^I \rightarrow s' \in C^I\}$	$\forall hasChild.Doctor$
existential	$\exists r.C$	$\{s \mid \exists s'.(s,s') \in r^I \wedge s' \in C^I\}$	$\exists hasChild.Lawyer$
minCard	$(\geqslant nr.C)$	$\{s \mid \sharp\{s'.(s,s') \in r^I \wedge s' \in C^I\} \geqslant n\}$	$(\geqslant 2 hasChild.Parent)$
maxCard	$(\leqslant nr.C)$	$\{s \mid \sharp\{s'.(s,s') \in r^I \wedge s' \in C^I\} \leqslant n\}$	$(\leqslant 1 hasChild.Parent)$
inverse	r^-	$\{(s,s') \mid (s,s') \in r^I\}$	$hasChild^-$

The terms ABox and TBox are used to describe two different types of statements in ontologies. TBox statements describe a system in terms of controlled vocabularies, e.g. a set of classes and properties. ABox are TBox-compliant statements about that vocabulary. Together ABox and TBox statements constitute a knowledge base.

Appendix B
Part of Experimental Data

Table B.1 Example of a service ontology

concept ticket
    provider **ofType** _string_
    trip **ofType** trip
    recordLocatorNumber **ofType** _integer_
concept _trip_
    origin **impliesType** _loc#location_
    destination **impliesType** _loc#location_
    departure **ofType** _date_
    arrival **ofType** _date_
concept tripFromAustria **subConceptOf** trip
    nonFunctionalProperties
        dc#relation **hasValue** tripFromAustriaDef
    endNonFunctionalProperties
axiom tripFromAustriaDef
    definedBy
        forall $\{?x, ?origin\}$(?x **memberOf** tripFromAustria
                implies?x[origin **hasValue** ?origin] **and**
                ?origin[_loc#locatedIn_ **hasValue** _loc#austria_])
concept reservationRequest
    nonFunctionalProperties
        dc#description **hasValue** "This concept represents a reservation request
                                for some trip for a particular person"
    endNonFunctionalProperties
    reservationItem **impliesType** _wsml#true_
    reservationHolder **impliesType** _prs#person_

Table B.2 *Turin_BelgiumChangeResidenceService*

---

namespace { _"http://www.semantic-gov.org/Italy#",
         wsmostudio _"http://www.wsmostudio.org#",
         gea _"http://www.semantic-gov.org/Ontologies/GEA#",
         dc _"http://purl.org/dc/elements/1.1#",
         wsmx _"http://www.wsmx.org/ontologies/oasm#",
         be _"http://www.semantic-gov.org/Belgium#"}
webService PA_Turin_BelgiumChangeResidenceService
   nonFunctionalProperties
      wsmostudio#version **hasValue** "0.7.3"
      _"http://www.semantic-gov.org#instanceOntology" **hasValue**
         PA_Turin_BelgiumChangeResidenceService_gea
   endNonFunctionalProperties
   importsOntology
   { _"http://www.semantic-gov.org/Belgium#GeaInstancesForPAServices",
   _"http://www.semantic-gov.org/Ontologies/GEA#GEA"}
   capability Turin_BelgiumChangeResidenceServiceCapability
      importsOntology
         { _"http://www.semantic-gov.org/Italy#TurinService"}
   sharedVariables ?request
   precondition
      definedBy
      ?person[hasCitizenship **hasValue** ?cit] **memberOf** Person
      and ?cit = belgianCitizenship.
   assumption
      definedBy
      ?request[document **hasValue** ?doc]
      and ?request[person **hasValue** ?person]
      and hasIdentityDocument(?person,?doc).
   postcondition
      definedBy
      ?request **memberOf** ChangeResidenceRequest
      and ?response **memberOf** ChangeResidenceResponse.
   effect
      definedBy
      ?response[message **hasValue** "Success"]
      and ?request[person **hasValue** ?person]
      and ?request[address **hasValue** ?address]
      and ?response[residenceCertificate **hasValue** ?certificate] **implies**
      lives(?person,?address) **and** hasDocument(?person,?certificate).

Table B.3 *geaTurin_BelgiumChangeResidenceService*

---

namespace {_"http://www.semantic-gov.org/Italy#",
             wsmostudio _"http://www.wsmostudio.org#",
             gea _"http://www.semantic-gov.org/Ontologies/GEA#",
             be _"http://www.semantic-gov.org/Belgium}
ontology PA_Turin_BelgiumChangeResidenceService_gea
  nonFunctionalProperties
     wsmostudio#version **hasValue** "0.7.3"
     _"http://www.semantic-gov.org#hasWsmoService" **hasValue**
     PA_Turin_BelgiumChangeResidenceService
  endNonFunctionalProperties
  importsOntology
    {_"http://www.semantic-gov.org/Belgium#GeaInstancesForPAServices",
    _"http://www.semantic-gov.org/Ontologies/GEA#GEA"}
  instance _# **memberOf** be#ChangeResidenceService
    gea#hasPADomain **hasValue** be#CitizenResidence
    gea#hasEffectType **hasValue** be#UpdateResidenceInformation
    gea#hasAdministrationLevel **hasValue** be#European
    gea#hasLocation **hasValue** {be#PAServiceEntryPoint, be#NotAvailable}
    gea#hasPublicServiceType **hasValue** be#EuropeanAdministrationService
    gea#isGovernedByLaw **hasValue** be#Lawn1228

Table B.4 Ontology *travel*1 from ISWC05 tutorial

---

wsmlVariant _"http://www.wsmo.org/wsml/wsml-syntax/wsml-flight"
namespace {_"http://deri.org/iswc2005tutorial/ontologies/travel1#",
             wsml _"http://www.wsmo.org/wsml/wsml-syntax#"}
ontology travel1
concept ticket
    type **ofType** _string
    departure_city **ofType** _string
    departure_code **ofType** _string
    arrival_city **ofType** _string
    arrival_code **ofType** _string
    departure_date **ofType** date
    arrival_date **ofType** date
    departure_time **ofType** time
    arrival_time **ofType** time
    issuing_terms **ofType** terms
    firstName **ofType** _string
    lastName **ofType** _string

Table B.5 Continuation of Table B.4

---

concept date
    year **ofType** _integer
    month **ofType** _integer
    day **ofType** _integer
concept time
    hour **ofType** _integer
    minutes **ofType** _integer
concept terms
    price **ofType** cost
    paymant_method **ofType** _string
    delivery_type **ofType** delivery
concept cost
    amount **ofType** _integer
    hasCurrency **ofType** currency
concept currency
instance euro **memberOf** currency
instance dollar **memberOf** currency
concept delivery
    type **ofType** _string
    due _to **ofType** date
concept customer
    firstName **ofType** _string
    lastName **ofType** _string
    street **ofType** _string
    city **ofType** _string
    zipCode **ofType** _string
    country **ofType** _string
concept trainTimeTable
    departure_city **ofType** _string
    arrival_city **ofType** _string
    travel_date **ofType** date
concept person
    name **ofType** _string
    age **ofType** _integer
    hasGender **ofType** gender
    hasChild **ofType** person
    marriedTo **ofType** person
concept gender
    value **ofType** _string
instance male **memberOf** gender
    value **hasValue** "male"
instance female **memberOf** gender
    value **hasValue** "female"

Table B.6 Ontology *travel*2 from ISWC05 tutorial

wsmlVariant _"http://www.wsmo.org/wsml/wsml-syntax/wsml-flight"
namespace {_"http://deri.org/iswc2005tutorial/ontologies/travel2#",
        wsml _"http://www.wsmo.org/wsml/wsml-syntax#"}
ontology travel2
concept date
    year **ofType** _integer
    month **ofType** _integer
    day **ofType** _integer
concept time
    hour **ofType** _integer
    minutes **ofType** _integer
concept payment
    amount **ofType** _integer
    inEuro **ofType** _boolean
    inDollars **ofType** _boolean
concept tripPoints
    departure **ofType** station
    arrival **ofType** station
concept station
    city **ofType** _string
    stationCode **ofType** _string
concept name
    first **ofType** _string
    last **ofType** _string
concept travelVoucher
    type **ofType** _string
    bearer **ofType** name
    toFrom **ofType** tripPoints
    departureDate **ofType** date
    arrivalDate **ofType** date
    departureTime **ofType** time
    arrivalTime **ofType** time
    terms **ofType** payment
    deliveryDate **ofType** date
concept human
    name **ofType** _string
    age **ofType** _integer
    noOfChildren **ofType** _integer
concept man **subConceptOf** human
    name **ofType** _string
    age **ofType** _integer
    noOfChildren **ofType** _integer
concept woman **subconceptOf** human
    name **ofType** _string
    age **ofType** _integer
    noOfChildren **ofType** _integer
concept marriage
    hasParticipant **ofType** human
    date **ofType** _date

Table B.7 Part of 115_Zip_Code_Lookup_Concepts.owl

```
<?xml version='1.0' encoding='ISO-8859-1'?>
     ...
<owl:Class rdf:ID="ArrayOfString_46350">
    <rdfs:subClassOf rdf:resource="&owl;#Thing"/>
</owl:Class>
<owl:DatatypeProperty rdf:ID="string_46353">
    <rdfs:range rdf:resource="http://www.w3.org/2001/XMLSchema#string"/>
    <rdfs:domain rdf:resource="#ArrayOfString_46350"/>
    <rdfs:subPropertyOf rdf:resource="null"/>
</owl:DatatypeProperty>
<owl:Class rdf:ID="GetZipByCityStateResponse_46351">
    <rdfs:subClassOf
        rdf:resource="http://moguntia.ucd.ie/daml/Datatypes.daml#ZIP_Code"/>
</owl:Class>
<owl:ObjectProperty rdf:ID="GetZipByCityStateResult_46354">
    <rdfs:range rdf:resource="#ArrayOfString_46350"/>
    <rdfs:domain rdf:resource="#GetZipByCityStateResponse_46351"/>
    <rdfs:subPropertyOf
        rdf:resource="http://moguntia.ucd.ie/daml/Datatypes.daml#ZIP_Code"/>
</owl:ObjectProperty>
<owl:Class rdf:ID="GetZipByCityState_46352">
    <rdfs:subClassOf rdf:resource="&owl;#Thing"/>
</owl:Class>
<owl:DatatypeProperty rdf:ID="CityName_46355">
    <rdfs:range rdf:resource="http://www.w3.org/2001/XMLSchema#string"/>
    <rdfs:domain rdf:resource="#GetZipByCityState_46352"/>
    <rdfs:subPropertyOf
        rdf:resource="http://moguntia.ucd.ie/daml/Datatypes.daml#City_Name"/>
</owl:DatatypeProperty>
<owl:DatatypeProperty rdf:ID="StateCode_46356">
    <rdfs:range rdf:resource="http://www.w3.org/2001/XMLSchema#string"/>
    <rdfs:domain rdf:resource="#GetZipByCityState_46352"/>
  <rdfs:subPropertyOf
        rdf:resource="http://moguntia.ucd.ie/daml/Datatypes.daml#State_Area_Code"/>
</owl:DatatypeProperty>
</rdf:RDF>
```

Table B.8 Service Ontology of 115_Zip_Code_Lookup.wsml

wsmlVariant _"http://www.wsmo.org/wsml/wsml-syntax/wsml-full"
namespace {_"http://wsao.deri.ie/Onto#"}
ontology _"http://wsao.deri.ie/Onto115_Zip_Code_Lookup"

concept ArrayOfString

concept GetZipByCityStateResponse **subconceptOf** Zip_code
concept GetZipByCityState
      cityName **ofType** string
      stateCode **ofType** string
concept Zip_code

Table B.9 Part of 115_Zip_Code_Lookup_Service.owl

```
<?xml version='1.0' encoding='ISO-8859-1'?>
    ...
<profileHierarchy:ZIP_Code_Resolver rdf:ID="Profile_Zip_Code_Lookup">
    <service:presentedBy rdf:resource="&the_service;"/ >
    <profile:has_process rdf:resource="&the_process;"/ >
    <profile:serviceName>
            Zip_Code_Lookup
    </profile:serviceName>
    <profile:textDescription>
            Gets the Zip Code given a city and state
    </profile:textDescription>
    <profile:hasInput rdf:resource="&the_concepts;#parameters_46360"/>
    <profile:hasOutput rdf:resource="&the_concepts;#parameters_46362"/>
    <profile:hasInput rdf:resource="&the_concepts;#CityName_46366"/>
    <profile:hasInput rdf:resource="&the_concepts;#StateCode_46367"/>
    <profile:hasOutput rdf:resource="&the_concepts;#Body_46369"/>
    <profile:hasInput rdf:resource="&the_concepts;#CityName_46373"/>
    <profile:hasInput rdf:resource="&the_concepts;#StateCode_46374"/>
    <profile:hasOutput rdf:resource="&the_concepts;#Body_46376"/>
</profileHierarchy:ZIP_Code_Resolver>
</rdf:RDF>
```

References

1. de Aalst, W.V.: Don't go with the flow: Web services composition standards exposed. IEEE Intelligent Systems 18(1), 72–76 (2003)
2. Agirre, E., Rigau, G.: A proposal for word sense disambiguation using conceptual distance. In: Proc. Intl. Conf. on Recent Advances in Natural Language Processing (1995)
3. Agirre, E., Rigau, G.: Word sense disambiguation using conceptual density. In: Proc. 16th Intl. Conf. on Computional Linguistics, pp. 16–22 (1996)
4. Akkiraju, R., Farrell, J., Miller, J., Nagarajan, M., Schmidt, M., Sheth, A., Verma, K.: Web service semantics — WSDL-S. A joint UGA-IBM Technical Note, version 1.0 (2005)
5. Andreopoulos, B., Alexopoulou, D., Schroeder, M.: Word sense disambiguation in biomedical ontologies with term co-occurrence analysis and document clustering. Intl. J. of Data Mining and Bioinformatics 2(3), 193–215 (2008)
6. Arpinar, I.B., Giriloganathan, K., Aleman-Meza, B.: Ontology quality by detection of conflicts in metadata. In: Proc. 4th Intl. EON Workshop: Evaluation of Ontologies for the Web (2006)
7. Baader, F., Calvanese, D., Nardi, D., Patel-Schneider, P.F. (eds.): The Description Login Handbook. Cambridge University Press (2003)
8. Banaei-Kashani, F., Chen, C.C., Shahbi, C.: WSPDS: Web services peer-to-peer discovery service. In: Proc. Intl. Symp. on Web Services and Applications (2004)
9. Banek, M., Vrdoljak, B., Tjoa, A.M.: Word sense disambiguation as the primary step of ontology integration. In: Proc. 19th Intl. Conf. on Database and Expert Systems Applications, pp. 65–72 (2008)
10. Banerjee, S.B.: An adapted lesk algorithm for word sense disambiguation using Word-Net. In: Proc. 3rd Intl. Conf. on Intelligent Text Processing and Computational Linguistics, pp. 17–22 (2002)
11. Batra, S., Bawa, S.: Semantic categorization of web services. Intl. J. of Recent Trends in Engineering 2(3), 20–23 (2009)
12. Batra, S., Bawa, S.: Semantic discovery of web services using principal component analysis. Intl. J. of the Physical Sciences 69(18), 4466–4472 (2011)
13. Benatallah, B., Hacid, M.-S., Rey, C., Toumani, F.: Request Rewriting-Based Web Service Discovery. In: Fensel, D., Sycara, K., Mylopoulos, J. (eds.) ISWC 2003. LNCS, vol. 2870, pp. 242–257. Springer, Heidelberg (2003)
14. Berners-Lee, T., Hendler, J., Lassila, O.: The semantic web. Scientific American 284(5), 34–43 (2001)

15. Bhuvaneswari, A., Karpagam, G.R.: Discovering substitutable and composable semantic web services for web service composition. Intl. J. of Computer Applications 48(8), 1–8 (2012)
16. Billiet, M., Bonomi, S., van der Graaf, E., Deiro, A., Kirgiannakis, E., Loutas, N., van Overeem, A., Peristeras, V., Papadopoulou, D., Savvas, I.: Identification and analysis of the SemanticGov showcase. Technical report, SemanticGov deliverable D2.3 (2006)
17. Biplav, S., Jana, K.: Web service composition — current solutions and open problems. In: Proc. Intl. Conf. on Automated Planning and Scheduling (2003)
18. Bisson, G., Nedellec, C., Canamero, L.: Designing clustering methods for ontology building — the mo'k workbench. In: Proc. ECAI Ontology Learning Workshop, pp. 13–19 (2000)
19. Bocchi, L., Frantechi, A., Goenczy, L., Koch, N.: Sensoria ontology: Prototype language for service modelling — ontology for SOAs presented through structured natural language. Technical report, Sensoria deliverable 1.1a (2006)
20. Borgida, A., Patel-Schneider, P.F.: A semantics and complete algorithm for subsumption in the classic description logic. J. of Artificial Intelligence Research 1, 277–308 (1994)
21. Bouchiha, D., Malki, M.: Semantic annotation of web services. In: Proc. Intl. Conf. on Web and Information Technologies, pp. 60–69 (2012)
22. Box, D., Ehnebuske, D., Kakivaya, G., Mendelsohn, N., Layman, A., Nielsen, H.F., Winer, D., Thatte, S.: Simple object access protocol (SOAP) 1.1. Technical report, W3C recommendation (2000)
23. Bray, T., Paoli, J., Sperberg-McQuee, C.M., Maler, E.: Extensible mark-up language (XML) 1.0 (2nd ed.). Technical report, W3C recommendation (2000)
24. Bruijn, J., Bussler, C., Domingue, J., Fensel, D.: Web service modeling ontology (WSMO). Technical report, WSMO final draft D2v1.2 (2005)
25. Budanitsky, A., Hirst, G.: Semantic distance in WordNet: An experimental, application-oriented evaluation of five measures. In: Proc. Workshop on WordNet and Other Lexical Resources (2001)
26. Buitelaar, P., Olejnik, D., Sintek, M.: Ontolt: A Protégé plug-in for ontology extraction from text. In: Proc. Intl. Semantic Web Conf. (2003)
27. Cabral, L., Domingue, J., Motta, E., Payne, T.R., Hakimpour, F.: Approaches to Semantic Web Services: an Overview and Comparisons. In: Bussler, C.J., Davies, J., Fensel, D., Studer, R. (eds.) ESWS 2004. LNCS, vol. 3053, pp. 225–239. Springer, Heidelberg (2004)
28. Chabeb, Y., Tata, S.: Yet another semantic annotation for WSDL. In: Proc. IADIS WWW/Internet Conf., pp. 462–467 (2008)
29. Chabeb, Y., Tata, S., Ozanne, A.: YASA-M: A semantic web service matchmaker. In: Proc. 24th IEEE Intl. Conf. on Advanced Information Networking and Applications, pp. 966–973 (2010)
30. Chalupsky, H.: Ontomorph: A translation system for symbolic knowledge. In: Proc. 7th Intl. Conf. on Knowledge Representation and Reasoning, pp. 471–482 (2000)
31. Channa, N., Li, S., Shi, W., Peng, G.: A CAN-based P2P infrastructure for semantic web services publishing and discovery. In: Proc. 1st IEEE and IFIP Intl. Conf. in Central Asia on Internet, p. 5 (2005)
32. Cho, I., McGregor, J., Krause, L.: A protocol-based approach to specifying interoperability between objects. In: Proc. 26th Technology of Object-Oriented Languages and Systems, pp. 84–96 (1998)
33. Christensen, E., Curbera, F., Meredith, G., Weerawarana, S.: Web services description language (WSDL) 1.1. Technical report, W3C recommendation (2001)

34. Cimiano, P., Völker, J.: Text2onto — a Framework for Ontology Learning and Data-Driven Change Discovery. In: Montoyo, A., Muñoz, R., Métais, E. (eds.) NLDB 2005. LNCS, vol. 3513, pp. 227–238. Springer, Heidelberg (2005)
35. Cornelis, C., De Kesel, P., Kerre, E.E.: Shortest paths in fuzzy weighted graphs. Intelligent and Soft Computing Techniques for Information Processing 19(11), 1051–1068 (2004)
36. Cost, S., Salzberg, S.: A weighted nearest-neighbour algorithm for learning with symbolic features. Machine Learning 10, 57–78 (1993)
37. Dan, B., Guha, R.V.: RDF vocabulary description language 1.0: RDF schema. Technical report, W3C recommendation (2004)
38. Deborah, L., Harmelen, F.: OWL web ontology language overview. Technical report, W3C recommendation (2004)
39. Dellschaft, K., Staab, S.: On how to perform a gold standard based evaluation of ontology learning. In: Proc. 5th Intl. Semantic Web Conf., pp. 228–241 (2006)
40. Dietrich, J.: The mandarax manual. Technical report, Institute of Information Sciences and Technology, New Zealand (2003)
41. Do, H.H., Rahm, E.: Coma — a system for flexible combination of schema matching approaches. In: Proc. 28th Intl. Conf. on Very Large Data Bases (2002)
42. Doan, A., Domingos, P., Halevy, A.: Reconciling schemas of disparate data sources: a machine learning approach. In: Proc. ACM SIGMOD (2001)
43. Domingue, J., Fensel, D., Hendler, J.A.: Handbook of semantic web technologies. Springer, Berlin (2011)
44. Dong, X., Halevy, A.Y., Madhavan, J., Nemes, E., Zhang, J.: Similarity search for web services. In: Proc. 30th Intl. Conf. on Very Large, pp. 372–383 (2004)
45. Ehrig, M., Euzenat, J.: State fo the art on ontology alignment. Technical report, KnowledgeWeb deliverable 2.2.3 (2004)
46. Ehrig, M., Haase, P., Hefke, M., Stojanovic, N.: Similarity for ontologies — a comprehensive framework. In: Proc. 13th European Conf. on Information Systems (2005)
47. Faloutsos, C., Oard, D.W.: A survey of information retrieval and filtering methods. Technical report CS-TR-3514, University of Maryland (1995)
48. Farrell, J., Lausen, H.: Semantic annotations for WSDL and XML Schema. Technical report, W3C recommendation 28 (2007)
49. Faure, D., Nedellec, C.: A corpus-based conceptual clustering method for verb frames and ontology. In: Proc. LREC Workshop on Adapting Lexical and Corpus Resources to Sublanguages and Applications, pp. 5–12 (1998)
50. Fayyad, U.M.: Advances in Knowledge Discovery and Data Mining. MIT Press, Cambridge (1996)
51. Fensel, D., Bussler, C.: The web service modeling framework WSMF. Electronic Commerce Research and Applications 1(2), 113–137 (2002)
52. Fensel, D., Keller, U., Lausen, H., Polleres, A., Toma, I.: WWW or what is wrong with web service discovery. In: Proc. Workshop on Frameworks for Semantics in Web Services (2005)
53. Fensel, D., Kerrigan, M., Zaremba, M.: Implementing semantic web services: The SESA framework. Springer, Berlin (2008)
54. Garcia, J.M., Ruiz, D., Ruiz-Cortes, A.: Improving semantic web services discovery using SPARQL-based repository filtering. J. of Web Semantics (2012)
55. Garlapati, S.S.: A comparison of SAWSDL based semantic web service discovery algorithms. Doctoral thesis, University of Georgia (2010)
56. De Giacomo, G., Lenzerini, M.: Tbox and Abox reasoning in expressive Description Logics. In: Proc. 5th Intl. Conf. on Principles of Knowledge Representation and Reasoning, pp. 316–327 (1996)

57. Ginsberg, A.: A unified approach to automatic index and information retrieval. IEEE Expert 8(5), 46–56 (1998)

58. Gomez-Perez, A., Corcho, O., Fernandez-Lopez, M.: Ontological engineering: with examples from the areas of knowledge management, e-commerce and the semantic web. Springer, New York (2004)

59. Gonzalez-Castillo, J., Trastour, D., Bartolini, C.: Description Logics for matchmaking of services. In: Proc. Workshop on Application of Description Logics (2001)

60. Gruber, T.R.: Toward principles for the design of ontologies used for knowledge sharing. Intl. J. of Human-Computer Studies 43(5/6), 907–928 (1995)

61. Guarino, N.: Formal ontology and information systems. In: Proc. 1st Intl. Conf. on Formal Ontologies in Information Systems, pp. 3–15 (1998)

62. Guarino, N., Welty, C.A.: An overview of Ontoclean. In: Handbook on Ontologies, pp. 151–172 (2004)

63. Guo, G., Yu, F., Chen, Z., Xie, D.: A method for semantic web service selection based on QoS ontology. J. of Computers 6(2), 377–386 (2011)

64. Haarslev, V., Moeller, R.: Racer: An OWL reasoning agent for the semantic web. In: Proc. Intl. Workshop on Applications, Products and Services of Web-based Support Systems, pp. 91–95 (2003)

65. Hacid, M., Leger, A., Rey, C., Toumani, F.: Dynamic discovery of e-services: a Description Logics based approach. In: Proc. 18th French Conf. on Advanced Databases, pp. 21–25 (2002)

66. Hahn, J., Schulz, S.: Towards very large terminological knowledge bases: A case study from medicine. In: Advances in AI, pp. 176–186. Springer, Berlin (2000)

67. Hau, J., Lee, W., Darlington, J.: A semantic similarity measure for semantic web services. In: Proc. 14th Intl. World Wide Web Conf. (2005)

68. Hearst, M.A.: Automatic acquisition of hyponyms from large text corpora. In: Proc. 15th Intl. Conf. on Computational Linguistic (1992)

69. Hoscheck, W.: The web service discovery architecture. In: Proc. IEEE/ACM Supercomputing Conf. (2002)

70. Hou, J., Zhang, J., Nayak, R., Bose, A.: Semantics-Based Web Service Discovery Using Information Retrieval Techniques. In: Geva, S., Kamps, J., Schenkel, R., Trotman, A. (eds.) INEX 2010. LNCS, vol. 6932, pp. 336–346. Springer, Heidelberg (2011)

71. Hovy, E.: Combining and standardizing large-scale, practical ontologies for machine translation and other uses. In: Proc. 1st Intl. Conf. on Language Resources and Evaluation, pp. 535–542 (1998)

72. Iqbal, K., Sbodio, M.L., Peristeras, V., Giuliani, G.: Semantic service discovery using SAWSDL and SPARQL. In: Proc. 4th Intl. Conf. on Semantics, Knowledge and Grid, pp. 205–212 (2008)

73. Jarmasz, M., Szpakowicz, S.: Roget's thesaurus and semantic similarity. In: Proc. Conf. on Recent Advances in Natural Language Processing, pp. 212–219 (2003)

74. Kanagwa, B., Lumaala, A.F.N.: Discovery of services based on WSDL tag level combination of distance measures. Intl. J. of Computing and ICT Research 6, 17–24 (2012)

75. Keskes, N., Lehireche, A., Rahmoun, A.: Web services selection based on context ontology and quality of services. Intl. Arab J. of e-Technology 1(3), 98–105 (2010)

76. Kietz, J.U., Maedche, A., Volz, R.: A method for semi-automatic ontology acquisition from a corporate intranet. In: Proc. Workshop Ontologies and Text (2000)

77. Kifer, M., Lara, R., Polleres, A., Zhao, C., Keller, U., Lausen, H., Fensel, D.: A logical framework for web service discovery. In: Proc. ISWC Workshop on Semantic Web Services (2004)

78. Kifer, M., Lausen, G.: F-logic: A higher-order language for reasoning about objects, inheritance and scheme. In: Proc. ACM SIGMOD, pp. 134–146 (1989)
79. Klusch, M., Fries, B., Sycara, K.: Automated semantic web service discovery with OWLS-MX. In: Proc. Joint Conf. on Autonomous Agents and Multiagent Systems (2006)
80. Klusch, M., Kapahnke, P., Zinnikus, I.: Hybrid adaptive web service selection with SAWSDL-MX and WSDL-Analyzer. In: Proc. 6th European Semantic Web Conf., pp. 550–564 (2009)
81. Kokash, N.: A comparison of web service interface similarity measures. In: Proc. European Starting AI Researcher Symp., pp. 220–231 (2006)
82. Kopecky, J., Vitvar, T., Bournez, C., Farrell, J.: SAWSDL: Semantic Annotations for WSDL and XML Schema. IEEE Internet Computing 11(6), 60–67 (2007)
83. Kreger, H.: Web services conceptual architecture. Technical report, IBM (2001)
84. Kuang, L., Deng, S.G., Li, Y., Shi, W., Wu, Z.H.: Exploring semantic technologies in service matchmaking. In: Proc. 3rd European Conf. on Web Services, pp. 226–234 (2005)
85. Kumar, S., Malik, S.K.: Towards semantic web based search engines. In: Proc. National Conf. on Advances in Computer Networks and Information Technology (2009)
86. Lara, R., Polleres, A., Lausen, H., Roman, D., de Bruijn, J., Fensel, D.: A conceptual comparison between WSMO and OWL. Technical report, WSMO deliverable D.4.1, v.0.1 (2005)
87. Lee, J.H., Kim, H., Lee, Y.J.: Information retrieval based on conceptual distance in is-a hierarchies. J. of Documentation 49, 188–207 (1993)
88. Lee, K., Jeon, J., Lee, W., Jeong, S., Park, S.: QoS for web services: Requirements and possible approaches. W3C Working Group Note 25 (2003)
89. Li, L., Horrocks, I.: A software framework for matchmaking based on semantic web technology. Intl. J. of Electronic Commerce 8(4), 39–60 (2004)
90. Li, X., Szpakowicz, S., Matwin, S.: A WordNet-based algorithm for word sense disambiguation. In: Proc. Intl. Joint Conf. on Artificial Intelligence, pp. 1368–1374 (1995)
91. Li, Y.H., Bandar, Z., McLean, D.: An approach for measuring semantic similarity between words using multiple information sources. IEEE Trans. Knowl. Data Eng. 15(4), 871–882 (2003)
92. Lin, D.: An information-theoretic definition of similarity. In: Proc. 15th Intl. Conf. on Machine Learning (1998)
93. Liu, Y., Ngu, H., Zeng, L.: QoS computation and policing in dynamic web service selection. In: Proc. 13th Intl. Conf. World Wide Web (2004)
94. Liu, Z., Guo, H.Q., Huang, Y.F.: Matchmaking for semantic web services based on OWL-S. In: Proc. 1st Intl. Conf. on Semantics, Knowledge and Grid, pp. 140–140 (2005)
95. Madhu, G., Govardhan, A., Rajinikanth, T.V.: Intelligent semantic web search engines: a brief survey. Intl. J. of Web & Semantic Technology 2(1), 34–42 (2011)
96. Maedche, A., Staab, S.: Mining Ontologies from Text. In: Dieng, R., Corby, O. (eds.) EKAW 2000. LNCS (LNAI), vol. 1937, pp. 189–202. Springer, Heidelberg (2000)
97. Maedche, A., Staab, S.: Semi-automatic engineering of ontologies from text. In: Proc. Intl. Conf. on Software Engineering and Knowledge Engineering (2000)
98. Maedche, A., Staab, S.: Ontology learning for the semantic web. IEEE Intelligent Systems 16(2), 72–79 (2001)
99. Maedche, A., Staab, S.: Measuring similarity between ontologies. In: Proc. European Conf. on Knowledge Acquisition and Management (2002)

100. Maedche, A., Staab, S.: Ontology learning. In: Handbook on Ontologies, pp. 173–189 (2004)

101. Makhlughian, M., Hashemi, S.M., Rastegari, Y., Pejman, E.: Web service selection based on ranking of QoS using associative classification. Intl. J. on Web Service Computing 3(1), 1–14 (2012)

102. Makhoul, J., Kubala, F., Schwartz, R., Weischedel, R.: Performance measures for information extraction. In: Proc. DARPA Broadcast News Workshop (1999)

103. Mani, A., Nagarajan, A.: Understanding quality of service for web services. Technical report, IBM Developerworks (2002)

104. Martin, D., Burstein, M., Hobbs, J., et al.: OWL-S: Semantic mark-up for web services. Technical report, W3C member submission (2004)

105. Maximilien, E.M., Singh, M.P.: A framework and ontology for dynamic web services selection. IEEE Internet Computing 8(5), 84–93 (2004)

106. McGuinness, D.L., Fikes, R., Rice, J., Wilder, S.: The Chimaera ontology environment. In: Proc. 17th National Conf. on Artificial Intelligence, pp. 1123–1124 (2000)

107. McIlraith, S.A., Son, T., Zeng, H.: Semantic web services. IEEE Intelligent Systems 16(2), 46–53 (2001)

108. McIlraith, S.A., Martin, D.L.: Bringing semantics to web services. IEEE Intelligent Systems 18(1), 90–93 (2003)

109. Menasce, D.A.: QoS issues in web services. IEEE Internet Computing 6(6), 72–75 (2002)

110. Michalsky, R.: Knowledge acquisition through conceptual clustering: a theoretical framework and algorithm for partitioning data into conjunctive concepts. Intl. J. of Policy Analysis and Information Systems 4(3), 219–243 (1980)

111. Miller, G., Beckwith, R., Fellbaum, C., Gross, D., Miller, K.: Introduction to WordNet: An online lexical database. Intl. J. of Lexicography 3(4), 235–244 (1990)

112. Miller, G.A., Leacock, C., Tengi, R., Bunker, R.: A semantic concordance. In: Proc. 13th Intl. Florida Artificial Intelligence Research Symp. Conf., pp. 219–223 (2000)

113. Morik, K., Kietz, J.U.: A bootstrapping approach to conceptual clustering. In: Proc. 6th Intl. Workshop on Machine Learning, pp. 503–504 (1989)

114. Motta, E., Domingue, J., Cabral, L., Feige, U.: IRS–II: A Framework and Infrastructure for Semantic Web Services. In: Fensel, D., Sycara, K., Mylopoulos, J. (eds.) ISWC 2003. LNCS, vol. 2870, pp. 306–318. Springer, Heidelberg (2003)

115. Mou, Y., Cao, J., Zhang, S.S., Zhang, J.H.: Interactive web service choice-making based on extended QoS model. In: Proc. 5th Intl. Conf. on Computer and Information Technology, pp. 1130–1134 (2005)

116. Muggleton, S., De Raedt, L.: Inductive logic programming: Theory and methods. J. of Logic Programming 19(20), 629–679 (1994)

117. Nardi, D., Brachman, R.J., Baader, F.: The Description Logic handbook: theory, implementation, and applications. Cambridge University Press (2003)

118. Navigli, R., Velardi, P., Cucchiarelli, A., Neri, F.: Extending and enriching WordNet with Ontolearn. In: Proc. 2nd Global WordNet Conf. (2004)

119. Nielsen, H.F., Gettys, J., Baird, S.A., Prudhommeaux, E., Lie, H.W., Lilley, C.: Network performance effects of http/1.1, css1, and png. ACM SIGCOMM Computer Communication Review 27(4), 155–166 (1997)

120. Noy, N., Musen, M.: Prompt: Algorithm and tool for automated ontology merging and alignment. In: Proc. 17th National Conf. on Artificial Intelligence (2000)

121. Oundhakar, S., Verma, K., Sivashanmugam, K., Sheth, A., Miller, J.: Discovery of web services in a multi-ontology and federated registry environment. Intl. J. of Web Services Research 1(3), 1–32 (2005)

122. Paik, I., Fujikawa, E.: Web service matchmaking using web search engine and machine learning. Intl. J. of Web Engineering 1(1), 1–5 (2012)

123. Pan, R., Ding, Z., Yu, Y., Peng, Y.: A Bayesian Network Approach to Ontology Mapping. In: Gil, Y., Motta, E., Benjamins, V.R., Musen, M.A. (eds.) ISWC 2005. LNCS, vol. 3729, pp. 563–577. Springer, Heidelberg (2005)

124. Paolucci, M., Kawamura, T., Payne, T.R., Sycara, K.: Semantic matching of web services capabilities. In: Horrocks, I., Hendler, J. (eds.) ISWC 2002. LNCS, vol. 2342, pp. 333–347. Springer, Heidelberg (2002)

125. Papaioannou, I.V., Tsesmetzis, D.T., Roussaki, I.G., Miltiades, E.A.: QoS ontology language for web services. In: Proc. 20th Intl. Conf. on Advanced Information Networking and Applications, vol. 1, pp. 101–106 (2006)

126. Papazoglou, M.P., Dubray, J.J.: A survey of web service technologies. Technical report DIT-04-058, University of Trento (2004)

127. Patel-Schneider, P.F., Fensel, D.: Layering the Semantic Web: Problems and Directions. In: Horrocks, I., Hendler, J. (eds.) ISWC 2002. LNCS, vol. 2342, pp. 16–29. Springer, Heidelberg (2002)

128. Patil, A., Oundhakar, S., Sheth, A., Verma, K.: Meteor-S web service annotation framework. In: Proc. 13th Intl. Conf. on World Wide Web, pp. 553–562 (2004)

129. Petrie, C., Bussler, C.: Service agents and virtual enterprises: A survey. IEEE Internet Computing 7(4), 68–78 (2003)

130. Petrie, C., Margaria, T., Lausen, H., Zaremba, M.: Semantic web services challenge: Results from the first year. Springer, New York (2008)

131. Pinheiro, P., Deborah, S., Mcguinness, L., Fikes, R.: A proof mark-up language for semantic web services. Information Systems 31(4), 381–395 (2006)

132. Plotkin, G.D.: A note on inductive generalisation. Machine Intelligence 5, 153–163 (1970)

133. Prud, E., Seaborne, A.: SPARQL query language for RDF. Technical report, W3C recommendation 15 (2008)

134. Purtilo, J., Atlee, J.M.: Module reuse by interface adaptation. Software — Practice and Experience 21(6), 539–556 (1991)

135. Rada, R., Mili, H., Bicknell, E., Blettner, M.: Development and application of a metric on semantic nets. IEEE Trans. on System, Man, and Cybernetics 19(1), 17–30 (1989)

136. Rahm, E., Bernstein, P.A.: A survery on approaches to automatic schema matching. VLDB Journal 10, 4 (2001)

137. Raj, J.R., Sasipraba, T.: Web service discovery based on computation of semantic similarity distance and QoS normalization. Indian J. of Computer Science and Engineering 3(2), 235–239 (2012)

138. Rajan, J.M., Lakshmi, M.D.: Ontology-based semantic search engine for healthcare services. Intl. J. on Computer Science and Engineering 4(4), 589–594 (2012)

139. Rajendran, T., Balasubramanie, P.: Analysis on the study of QoS-aware web services discovery. J. of Computing 1(1), 119–130 (2009)

140. Ram, S., Park, J.: Semantic conflict resolution ontology (SCROL): An ontology for detecting and resolving data and schema level conflicts. IEEE Trans. on Knowledge and Data Engineering 16(2), 189–202 (2004)

141. Ran, S.: A model for web services discovery with QoS. ACM SIGecom Exchanges 4(1), 1–10 (2003)

142. Rao, J., Su, X.: A Survey of Automated Web Service Composition Methods. In: Cardoso, J., Sheth, A.P. (eds.) SWSWPC 2004. LNCS, vol. 3387, pp. 43–54. Springer, Heidelberg (2005)

143. Ren, K., Chen, J., Chen, T., Song, J., Xiao, N.: Grid-based semantic web service dis-covery model with QoS constraints. In: Proc. rrd Intl. Conf. on Semantics, Knowledge and Grid, pp. 479–482 (2007)

144. Resnik, P.: Semantic similarity in a taxonomy: An information-based measure and its application to problems of ambiguity in natural language. J. of Artificial Intelligence Research 11, 95–130 (1999)

145. Ristad, E.S., Yianilos, P.N.: Learning string edit distance. IEEE Trans. on Pattern Anal-ysis and Machine Intelligence 20(5), 522–532 (1998)

146. Sabou, M., Wroe, C., Goble, C., Stuckenschmidt, H.: Learning domain ontologies for semantic web service descriptions. J. of Web Semantics 3(4), 340–365 (2005)

147. Salton, G.: Automatic text processing: The transformation, analysis and retrieval of information by computer. Addision-Wesley, Reading (1989)

148. Sapkota, B., Roman, D., Kruk, S.R., Fensel, D.: Distributed web service discov-ery architecture. In: Proc. Intl. Conf. on Internet and Web Applications and Ser-vices/Advanced Intl. Conf., pp. 136–136 (2006)

149. Sathya, M., Swarnamugi, M., Dhavachelvan, P., Sureshkumar, G.: Evaluation of QoS based web-service selection techniques for service composition. Intl. J. of Software Engineering 1(5), 73–90 (2011)

150. Sbodio, M.L., Martin, D., Moulin, C.: Discovering semantic web services using SPARQL and intelligent agents. J. of Web Semantics 8, 310–328 (2010)

151. Schmidt, C., Parashar, M.: A peer-to-peer approach to web service discovery. World Wide Web 7(2), 211–229 (2004)

152. Schulte, S., Lampe, U., Eckert, J., Steinmetz, R.: LOG4SWS.KOM: Self-adapting se-mantic web service discovery for SAWSDL. In: Proc. 6th World Congress on Services, pp. 511–518 (2010)

153. Sintek, M., Buitelaar, P., Olejnik, D.: A formalization of ontology learning from text. In: Proc. ISWC Workshop on Evaluation of Ontology-based Tools (2004)

154. Sirin, E., Parsia, B.: SPARQL-DL: SPARQL Query for OWL-DL. In: Proc. OWLED Workshop on OWL: Experiences and Directions (2007)

155. Steinmetz, N., de Bruijn, J., Frankl, A.: D37v0.2 WSML/OWL mapping. Technical report, WSML Working Draft (2008)

156. Steve, G., Gangemi, A., Pisanelli, D.M.: Integrating medical terminologies with onions methodology. In: Info. Modelling and Knowledge Bases VIII. IOS Press, Amsterdam (1998)

157. Stollberg, M.: Scalable semantic web service discovery for goal-driven service-oriented architectures. Doctoral thesis, Leopold-Franzens-Universität Innsbruck (2008)

158. Studer, R., Volz, R., Stumme, G., Hotho, A.: Semantic web — state of the art and future directions. KI Heft, Special Issue on the Semantic Web 3, 5–9 (2003)

159. Stumme, G., Maedche, A.: FCA-merge: Bottom-up merging of ontologies. In: Proc. 17th Intl. Conf. on Artificial Intelligence, pp. 225–230 (2001)

160. Suganyakala, R., Aarthilakshmi, M., Karpagam, G.R., Maheswari, S.: Ontology based QoS driven web service discovery. Intl. J. of Computer Science Issues 8(1), 191–198 (2011)

161. Sycara, K., Klusch, M., Wido, S., Lu, J.: Dynamic service matchmaking among agents in open information environments. ACM SIGMOD Rec. 28, 47–53 (1999)

162. Sycara, K., Widoff, S., Klusch, M., Lu, J.: Larks: Dynamic matchmaking among het-erogeneous software agents in cyberspace. J. of Autonomous Agents and Multi-Agent Systems 5(2), 173–203 (2002)

163. Tata, S., Patel, J.M.: Estimating the selectivity of TF-IDF based cosine similarity pred-icates. ACM SIGMOD Rec. 36(2), 7–12 (2007)

164. Tsesmetzis, D.T., Roussaki, I.G., Papaioannou, I.V., Anagnostou, M.E.: QoS awareness support in web-service semantics. In: Proc. Advanced Intl. Conf. on Telecommunications and Intl. Conf. on Internet and Web Applications and Services, p. 128 (2006)

165. UDDI.ORG: Introduction to UDDI: Import features and functional concepts. Technical report, OASIS OPEN (2004)

166. Tümer, D., Shah, M.A., Bitirim, Y.: An empirical evaluation on semantic search performance of keyword-based and semantic search engines: Google, Yahoo, MSN and Hakia. In: Proc. 4th Intl. Conf. on Internet Monitoring and Protection, pp. 51–55 (2009)

167. Van Rijsbergen, C.J.: Information Retrieval. University of Glasgow (1979)

168. Verma, K., Sivashanmugam, K., Sheth, A., Patil, A., Oundhakar, S., Miller, J.: Meteor-S WSDI: A scalable P2P infrastructure of registries for semantic publication and discovery of web services. J. of Information Technology and Management 6(1), 17–39 (2005)

169. Vitvar, T., Zaremba, M., Moran, M., Fensel, D.: SESA: Emerging technology for service-centric environments. IEEE Software 24(6), 56–67 (2007)

170. Volz, R., Oberle, D., Staab, S., Motik, B.: KAON server — a semantic web management system. In: Proc. 12th Intl. World Wide Web Conf. (2003)

171. Vu, L.H., Hauswirth, M., Aberer, K.: Towards P2P-based semantic web service discovery with QoS supports. In: Proc. Workshop on Business Processes and Services (2005)

172. Wang, H., Zhai, S., Fan, L.: Query for semantic web services using SPARQL-DL. In: Proc. 2nd Intl. Symp. on Knowledge Acquisition and Modeling, pp. 367–370 (2009)

173. Wang, T., Wei, D., Wang, J., Bernstein, A.: SAWSDL-iMatcher: a customizable and effective semantic web service matchmaker. J. of Web Semantics 9(4), 402–417 (2011)

174. Wang, X., Vitvar, T., Peristeras, V., Mocan, A., Goudos, S.K., Tarabanis, K.: WSMO-PA: Formal specification of public administration service model on semantic web service ontology. In: Proc. 40th Hawaii Intl. Conf. on Systems Sciences (2007)

175. Wang, X., Nazir, S., Loutas, N., Peristeras, V., Halang, W.A.: Ontology-dependent two-phase semantic web services discovery and its e-government implementation. Int. J. of Business Process Integration & Management 5(2), 173–184 (2011)

176. Wang, Y., Stroulia, E.: Flexible interface matching for web-service discovery. In: Proc. 4th Intl. Conf. on Web Information Systems Engineering (2003)

177. Wang, Y., Stroulia, E.: Semantic structure matching for assessing web-service similarity. In: Proc. 1st Intl. Conf. on Service Oriented Computing, pp. 194–207 (2004)

178. Wu, J., Chen, L., Xie, Y., Zheng, Z.: Titan: a system for effective web service discovery. In: Proc. WWW 2012 — Demos Track, pp. 441–444 (2012)

179. Wu, W., Doan, A., Yu, C.T., Meng, W.: Bootstrapping domain ontology for semantic web services from source web sites. In: Proc. VLDB 2005 Workshop on Technologies for E-Services, pp. 11–22 (2005)

180. Yang, Y., Pedersen, J.O.: A comparative study on feature selection in text categorization. In: Proc. Intl. Conf. on Machine Learning (1997)

181. Yates, R.B., Neto, B.R.: Modern information retrieval. ACM Press, New York (1999)

182. Zaremski, A.M., Wing, J.M.: Signature matching: a tool for using software libraries. ACM Trans. on Software Engineering and Methodology 4(2), 146–170 (1995)

183. Zaremski, A.M., Wing, J.M.: Specifications matching of software components. ACM Trans. on Software Engineering and Methodology 6(4), 333–369 (1997)

184. Zhang, Y., Zheng, Z., Lyu, M.R.: WSExpress: A QoS-aware search engine for web services. In: Proc. IEEE Intl. Conf. on Web Services, pp. 91–98 (2010)

185. Zhou, C., Chia, L.T., Lee, B.B.: DAML-QoS ontology for web services. In: Proc. Intl. Conf. on Web Services, pp. 472–479 (2004)